ACHIEVING HIGH-PERFORMANCE FEDERAL FACILITIES

STRATEGIES AND APPROACHES FOR TRANSFORMATIONAL CHANGE

Committee on High-Performance Green Federal Buildings: Strategies and Approaches for Meeting Federal Objectives

Board on Infrastructure and the Constructed Environment
Division on Engineering and Physical Sciences

NATIONAL RESEARCH COUNCIL
OF THE NATIONAL ACADEMIES

THE NATIONAL ACADEMIES PRESS
Washington, D.C.
www.nap.edu

THE NATIONAL ACADEMIES PRESS   500 Fifth Street, N.W.   Washington, DC 20001

NOTICE: The project that is the subject of this report was approved by the Governing Board of the National Research Council, whose members are drawn from the councils of the National Academy of Sciences, the National Academy of Engineering, and the Institute of Medicine. The members of the committee responsible for the report were chosen for their special competences and with regard for appropriate balance.

This report was primarily supported by Sponsor Award No. XW001-XW994 between the National Academy of Sciences and the U.S. General Services Administration. Any opinions, findings, conclusions, or recommendations expressed in this publication are those of the author(s) and do not necessarily reflect the views of the organizations or agencies that provided support for the project.

International Standard Book Number-13: 978-0-309-21168-0
International Standard Book Number-10: 0-309-21168-9

Additional copies of this report are available from the National Academies Press, 500 Fifth Street, N.W., Lockbox 285, Washington, DC 20055; (800) 624-6242 or (202) 334-3313 (in the Washington metropolitan area); Internet, http://www.nap.edu.

THE NATIONAL ACADEMIES
Advisers to the Nation on Science, Engineering, and Medicine

The **National Academy of Sciences** is a private, nonprofit, self-perpetuating society of distinguished scholars engaged in scientific and engineering research, dedicated to the furtherance of science and technology and to their use for the general welfare. Upon the authority of the charter granted to it by the Congress in 1863, the Academy has a mandate that requires it to advise the federal government on scientific and technical matters. Dr. Ralph J. Cicerone is president of the National Academy of Sciences.

The **National Academy of Engineering** was established in 1964, under the charter of the National Academy of Sciences, as a parallel organization of outstanding engineers. It is autonomous in its administration and in the selection of its members, sharing with the National Academy of Sciences the responsibility for advising the federal government. The National Academy of Engineering also sponsors engineering programs aimed at meeting national needs, encourages education and research, and recognizes the superior achievements of engineers. Dr. Charles M. Vest is president of the National Academy of Engineering.

The **Institute of Medicine** was established in 1970 by the National Academy of Sciences to secure the services of eminent members of appropriate professions in the examination of policy matters pertaining to the health of the public. The Institute acts under the responsibility given to the National Academy of Sciences by its congressional charter to be an adviser to the federal government and, upon its own initiative, to identify issues of medical care, research, and education. Dr. Harvey V. Fineberg is president of the Institute of Medicine.

The **National Research Council** was organized by the National Academy of Sciences in 1916 to associate the broad community of science and technology with the Academy's purposes of furthering knowledge and advising the federal government. Functioning in accordance with general policies determined by the Academy, the Council has become the principal operating agency of both the National Academy of Sciences and the National Academy of Engineering in providing services to the government, the public, and the scientific and engineering communities. The Council is administered jointly by both Academies and the Institute of Medicine. Dr. Ralph J. Cicerone and Dr. Charles M. Vest are chair and vice chair, respectively, of the National Research Council.

www.national-academies.org

Preface

Buckminster Fuller once said, "The best way to predict the future is to design it." If the United States is to meet the challenges of global climate change, energy security, and environmental sustainability, an essential element for doing so is the design and retrofit of buildings.

The numbers are well known. Buildings account for almost 40 percent of primary energy use in the United States, 12 percent of total water use, and 60 percent of all nonindustrial waste. In addition, the indoor environmental quality of buildings affects the health, safety, and productivity of the people who occupy them.

In recognition of these impacts, building design and management, building technologies, and tools for analysis and decision-support are evolving. Today it is possible to create "high-performance" buildings: buildings that are more environmentally sustainable, that support occupant health, safety, and productivity, and that are cost-effective throughout their life cycles.

The U.S. federal government has the opportunity, and the responsibility, to significantly improve the performance of its buildings and to lead the way for other large organizations to do the same. Today, the government owns or leases 429,000 buildings worldwide, containing 3.34 billion square feet of space. Congress and two presidential administrations have enacted legislation and issued executive orders aimed at transforming the existing portfolio of federal buildings into one of high-performance facilities. In addition to achieving significant environmental benefits, such a transformation will result in long-term reductions in operations, maintenance, and life-cycle costs. Given these factors, the question now is not "Why should the federal government develop high-performance buildings as a matter of course?" Instead, federal decision makers at all levels and in all agencies should be required to justify why they would continue to construct and retrofit buildings in conventional ways.

In 2010, the U.S. General Services Administration's (GSA's) Office of Federal High-Performance Green Buildings asked the National Academies to appoint an ad hoc committee of experts to conduct a public workshop and prepare a report that identifies strategies and approaches for achieving a range of objectives associated with federal high-performance green buildings.

The committee conducted the workshop in July 2010. The speakers included early adopters of transformational strategies for achieving a sustainable built environment. They identified regional, local, and international initiatives involving federal agencies, municipalities, states, and universities. More than 60 practitioners from the public, private, and nonprofit sectors and academia participated in the workshop and gave generously of their time and knowledge. The committee was especially impressed by the number of federal agencies represented at the workshop and their enthusiastic support for the subject. Their ideas and others are integrated into this report.

Chapter 1, "Context," focuses on trends in building design, operations, and management, provides statistics and other background information on federal facilities, and discusses the committee's approach for fulfilling its statement of task.

Chapter 2, "Objectives and Challenges Associated with Federal High-Performance Green Buildings," identifies many of the objectives related to federal high-performance green buildings as established by legislation, executive orders, and other mandates. Long-standing, well-documented challenges and obstacles that hinder achievement of the established objectives are also discussed.

Chapter 3, "Levers of Change," identifies areas where federal agencies can leverage their resources to spur transformational actions and make sustainability the preferred choice at all levels of decision making. The "levers of change" relate to all phases of buildings' life cycles and can be immediately used by federal agencies to overcome the identified challenges and barriers.

Chapter 4, "Best Practices, Tools, and Technologies for Transformational Change," highlights a range of practices, tools, and technologies identified at the public workshop and throughout the course of this activity. It highlights ways that federal agencies can achieve objectives associated with federal high-performance green buildings.

Chapter 5, "Strategies and Approaches for Achieving a Range of Objectives Associated with High-Performance Federal Facilities," synthesizes the committee's findings and conclusions from Chapters 1 to 4 into 12 wide-ranging strategies and approaches for achieving a range of objectives associated with high-performance green federal facilities.

Appendixes D through I contain write-ups of many of the presentations given to the committee that provide practical and inspirational examples for creating more sustainable facilities. Although some of these examples are highlighted in the main body of the report, readers are urged to delve into the appendixes for additional context and ideas.

The committee thanks the following people whose presentations are the source for many of the best practices cited in the report: Hal Alguire, Jeffrey Baker, Robert Berkebile, Peter Garforth, Thomas Hall, Christopher Juniper, Greg Kats, William Miner, Mark Mykleby, Greg Norris, David Orr, and Roland Risser.

The committee was impressed by the enlightened approach taken by GSA's Office of Federal High-Performance Green Buildings and thanks Kevin Kampschroer, Katherine "Joni" Teter, Michael Bloom, and Ken Sandler for their insights and assistance throughout. Additional leadership and valuable assistance were provided by Shyam Sunder, Dale Manty, and Paul Domich of the Building Technology Research and Development Subcommittee of the National Science and Technology Council and by Michelle Moore of the Office of the Federal Environmental Executive.

For me, it was an honor and a privilege to work with the other members of the committee. Each person was a recognized expert in his or her field. Each volunteered his or her time and expertise as a public service and melded a large and varied set of information together to produce this report.

As a group, the committee believes that the time is now to move forward aggressively to create a portfolio of high-performance federal facilities. Federal agencies have the required tools, technologies, and knowledge. Effective implementation requires conscientious, directional, and strategic decision making at every level of government. Success will require leadership, a willingness to use collaborative approaches to overcome conventional thinking, and sustained commitment over several decades. The result will be a higher quality of life and a higher-quality environment.

David J. Nash, *Chair*
Committee on High-Performance Green Federal Buildings:
Strategies and Approaches for Meeting Federal Objectives

Acknowledgment of Reviewers

The authoring committee acknowledges the significant contributions made by the workshop participants, all of whom willingly and enthusiastically volunteered their time and ideas.

This report has been reviewed in draft form by individuals chosen for their diverse perspectives and technical expertise, in accordance with procedures approved by the National Research Council's Report Review Committee. The purpose of this independent review is to provide candid and critical comments that will assist the institution in making its published report as sound as possible and to ensure that the report meets institutional standards for objectivity, evidence, and responsiveness to the study charge. The review comments and draft manuscript remain confidential to protect the integrity of the deliberative process. We wish to thank the following individuals for their review of this report:

Jonathan Barnett, University of Pennsylvania,
Carmine Battafarano, Burns and Roe Services Corporation,
Bill Browning, Terrapin/Bright Green LLC,
Michael Johnson, University of Arkansas,
Matthys Levy, Weidlinger Associates (emeritus),
Annie Pearce, Virginia Tech,
Chris Poland, Degenkolb Engineers,
Maxine Savitz, Honeywell Inc. (retired), and
Alan Shimada, ENVIRON, Inc.

Although the reviewers listed above have provided many constructive comments and suggestions, they were not asked to endorse the conclusions or recommendations, nor did they see the final draft of the report before its release. The review of this report was overseen by Richard N. Wright, National Institute of Standards and Technology (retired). Appointed by the National Research Council, he was responsible for making certain that an independent examination of this report was carried out in accordance with institutional procedures and that all review comments were carefully considered. Responsibility for the final content of this report rests entirely with the authoring committee and the institution.

Contents

Summary

The design, construction, operation, and retrofit of buildings is evolving in response to ever-increasing knowledge about the impact of indoor environments on people and the impact of buildings on the environment. Research has shown that the quality of indoor environments can affect the health, safety, and productivity of the people who occupy them. Buildings are also resource intensive, accounting for 40 percent of primary energy use in the United States, 12 percent of water consumption, and 60 percent of all nonindustrial waste. The processes for producing electricity at power plants and delivering it for use in buildings account for 40 percent of U.S. greenhouse gas emissions.

The scale of building design is also evolving. The focus has shifted from individual buildings to entire portfolios (groups of buildings under a single management), to neighborhoods, communities, regions, watersheds, airsheds, and economies. As the scale and scope of design have increased, the opportunities for sharing infrastructure, conserving land and open space, and preserving and regenerating environmental systems have also increased.

Greater knowledge about buildings and their impacts has led to new processes and tools for measuring and evaluating buildings' performance throughout their life cycles: planning, design, construction, operations, maintenance, retrofit, and deconstruction. New technologies are being developed that can help reduce greenhouse gas emissions and the use of fossil fuels, energy, and water and that provide electricity from renewable energy sources.

The U.S. federal government manages approximately 429,000 buildings of many types with a total square footage of 3.34 billion worldwide, of which about 80 percent is owned space. More than 30 individual departments and agencies are responsible for managing these buildings. The characteristics of each agency's portfolio of facilities are determined by the agency's mission and its programs.

Recognizing the significant role of buildings in solving national issues such as energy independence and security, global climate change, and environmental sustainability, and recognizing the opportunity for federal leadership, Congress and two presidential administrations have enacted laws and issued executive orders directing federal agencies to develop high-performance, energy-efficient, sustainable federal buildings. They include the Energy Independence and Security Act (EISA) of 2007, Executive Order 13423, *Strengthening Federal Environmental, Energy, and Transportation Management* (2007), and Executive Order 13514, *Federal Leadership in Environmental, Energy, and Economic Performance* (2009).

Together, these mandates establish more than 20 objectives for federal high-performance buildings. The objectives include reducing the use of energy, potable water, fossil fuels, and materials; reducing greenhouse gas emissions; improving indoor environmental quality; increasing the use of recycling and environmentally preferable products; minimizing waste and pollutants through source reduction; pursuing cost-effective innovative strategies to minimize consumption of energy, water, and materials; leveraging agency acquisitions to foster markets for sustainable technologies, materials, products, and services; locating new buildings in sustainable locations; participating in regional transportation planning; and strengthening the vitality and livability of the communities in which federal facilities are located. EISA requires agencies to eliminate fossil fuel energy use in new buildings and major renovations by 2030. Executive Order 13514 directs that beginning in 2020 and thereafter, all new federal buildings that enter the planning process should be designed to achieve zero-net-energy use by 2030.[1]

Each mandate specifically calls for the use of a life-cycle perspective or life-cycle costing, establishes interim and longer-term goals and objectives, and establishes baselines and performance measures for evaluating progress in achieving the goals. EISA also established the Office of Federal High-Performance Green Buildings within the U.S. General Services Administration (GSA).

STATEMENT OF TASK AND THE COMMITTEE'S APPROACH

In 2010, GSA's Office of Federal High-Performance Green Buildings asked the National Academies to appoint an ad hoc committee of experts to conduct a public workshop and prepare a report that identifies strategies and approaches for achieving a range of objectives associated with high-performance green federal buildings. To meet its charge, the committee was asked to identify the following:

- Challenges, barriers, and gaps in knowledge related to developing high-performance green federal buildings.
- Current best practices and ways to optimize resources for achieving high-performance green building objectives during planning, design, construction, operations, and maintenance for new and existing facilities.
- Best practices for reporting the outcomes of investments in high-performance green federal buildings in a transparent manner on public federal Web sites.
- Approaches, tools, and technologies for overcoming identified challenges, barriers, and gaps in knowledge.

The committee recognized up front that many other reports, papers, and books have been published and databases have been created related to various aspects of high-performance green buildings. In addition, many initiatives are under way within federal agencies and other public and private organizations, universities, nonprofit entities, and community groups, across the country and internationally. To try to capture all of the valuable and thought-provoking ideas, lessons learned, and evidence-based data from these initiatives would not be possible.

The committee determined it would focus on identifying examples of important initiatives taking place and available resources and on ascertaining how these examples and resources could be used to

---

[1]The executive order defines a zero-net-energy building as one that is designed, constructed, and operated to require a greatly reduced quantity of energy to operate, to meet the balance of energy needs from sources of energy that do not produce greenhouse gases, and to therefore result in no net emissions of greenhouse gases and be economically viable.

help make sustainability the preferred choice at all levels of decision making. In this way, the report could also be of value to many federal agencies with differing missions, types of facilities, and operating procedures: It is up to the individual agencies to adapt the approaches to their situations.

The committee held the public workshop on July 20 and 21, 2010, met two additional times, and corresponded through conference calls and e-mail. Information was provided by federal staff from the GSA and other federal agencies; representatives of organizations that have achieved significant break-throughs in developing high-performance green buildings, installations, campuses, and communities; and more than 60 practitioners from public agencies, industry, and academia.

CHALLENGES AND BARRIERS TO ACHIEVING OBJECTIVES ASSOCIATED WITH FEDERAL HIGH-PERFORMANCE GREEN BUILDINGS

Federal agencies will need to overcome a number of challenges and barriers if they are to achieve the goals and objectives associated with high-performance green buildings. The challenges include these:

- Embedding sustainability into everyday decision making.
- Excess facilities that siphon off already constrained resources.
- The federal budget process.
- Segmented processes that fail to optimize resources.
- Lack of alignment between reporting requirements and performance measurement systems.
- Perceived higher costs of building green.
- Workforce skills and training.
- Widespread deployment of innovative technologies for high-performance buildings.
- Gaps in knowledge on a range of topics and technologies.

LEVERS OF CHANGE AND BEST PRACTICES

To overcome these challenges and barriers, the committee identified "levers of change," which it defined as "areas where federal agencies can leverage their resources to spur transformational actions and to make sustainability the preferred choice at all levels of decision making." The levers support an overall life-cycle perspective and can be used in all phases of building design, operation, and management. They represent changes in mindset as much as changes in processes or technologies. The levers are the following:

- Systems-based thinking,
- Portfolio-based facilities management,
- Integrated work processes,
- Procurement, contracting, and finance,
- Communication and feedback for behavioral change,
- Standards and guidelines, and
- Technologies and tools.

For its report, the committee defined best practices as "processes, procedures, or technologies that aim to optimize available resources and that could be effectively applied by the GSA and other federal agencies to meet similar objectives." The definition is intentionally broad, because new practices, tech-nologies, tools, and processes related to high-performance green buildings are continually emerging.

The committee believed that agencies may lose opportunities to leapfrog ahead to fulfill their mandates, if the committee only recommended well-documented best practices with a history of proven results.

STRATEGIES AND APPROACHES FOR ACHIEVING A RANGE OF OBJECTIVES ASSOCIATED WITH FEDERAL HIGH-PERFORMANCE FACILITIES

The committee identified 12 strategies and approaches that the GSA and all federal agencies can use to achieve a range of objectives associated with high-performance buildings and facilities. They are based on the levers of change, on the best practices, tools, and technologies identified at the public workshop and other meetings, and on the committee members' own expertise. They can be applied to portfolios of facilities and to individual building projects.

The strategies and approaches are summarized in Box S.1. Brief explanations for each follow. More detailed information, including examples of best practices, tools, and technologies for implementing the strategies and approaches, is provided in Chapter 5.

BOX S.1
Summary of Strategies and Approaches for Achieving a Range of Objectives Related to Federal High-Performance Facilities

1. Use systems-based thinking and life-cycle assessment to identify new ways to provide services and to eliminate waste.
2. Focus on community- and regional-based approaches to fill gaps, leverage resources, and optimize results.
3. Align existing federal facilities to current missions and consolidate the total facilities footprint to lower costs, reduce carbon emissions, reduce water and energy use, and optimize available resources.
4. Operate facilities efficiently to optimize their performance.
5. Aggressively implement proven sustainable technologies as a matter of course.
6. Use integrated, collaborative processes and practices to overcome conventional segmented processes that fail to optimize resources.
7. Aim for high-performance, near-zero-net-energy buildings now.
8. Measure, verify, and report performance to improve processes and change behavior.
9. Use performance-based approaches to unleash the creativity of contractors.
10. Collaborate to drive the market for sustainable products and high-performance technologies.
11. Use standards and guidelines to drive change and embed sustainability into decision-making processes.
12. Communicate successes and learn from others.

1. Use systems-based thinking and life-cycle assessment to identify new ways to provide services and to eliminate waste.

Systems-based thinking provides a life-cycle perspective that can overcome challenges posed by the federal budget process and by segmented work processes. As importantly, it can help federal agencies identify new ways to use resources, to substitute more sustainable resources, to eliminate waste, and to avoid narrowly focused solutions with unintended consequences.

Systems-based thinking begins with the development of goals and objectives for the activity: The more ambitious the goals, the more innovative the strategies are likely to be. A systems-based approach can be especially effective in helping federal agencies meet their goals for reducing greenhouse gas emissions, reducing the use of potable water, conserving and protecting water resources, for recycling and pollution prevention, for minimizing the generation of waste and pollutants through source reduction, and for regional transportation planning.

2. Focus on community- and regional-based approaches to fill gaps, leverage resources, and optimize results.

Where federal facilities occupy large, contiguous land areas, such as military bases, research campuses, office parks, embassy compounds, and the like, they have opportunities to save energy, reduce the use of fossil fuels, and reduce greenhouse gas emissions by building on-site combined heat and power (co-generation) plants, installing solar arrays and wind turbines for on-site generation of renewable energy, and installing district energy systems and other technologies. Larger-scale development also facilitates the recycling of potable water and stormwater management.

Most federal facilities are dependent, in part, on nonfederal infrastructure systems for power, water, wastewater removal, transportation, and telecommunications. Federal agencies can leverage their available resources and achieve goals for strengthening the vitality and livability of adjacent communities by forming partnerships with local communities, utility companies, and others with shared interests.

3. Align existing federal facilities to current missions and consolidate the total facilities footprint to lower costs, reduce carbon emissions, reduce water and energy use, and optimize available resources.

Effective portfolio-based facilities management optimizes the performance of existing buildings and other facilities in support of an organization's mission, carefully considers the addition and location of new buildings, and uses life-cycle costing for all potential investments. Federal agencies can use portfolio-based management to align their facilities with mission; to determine which facilities are excess; to identify noncapital solutions for providing required services and avoid the long-term costs and environmental impacts of new buildings; to choose sustainable locations for new buildings; to determine where space can be consolidated; and to optimize the performance of existing buildings. In these ways, effective portfolio-based facilities management can help agencies meet an array of environmental and cost objectives for high-performance facilities.

To effectively implement a portfolio-based facilities management approach, federal agencies need a well-trained workforce. The Federal Buildings Personnel Training Act of 2010, when implemented, should help federal managers strengthen the skills of their workforces for operating high-performance green buildings and for portfolio-based facilities management.

4. Operate facilities efficiently to optimize their performance.

The vast majority of facilities that federal agencies will be using in 2020, 2030, and 2040 exist today. Operating building systems as they were designed can result in significant reductions in the consumption of energy and water, and can contribute positively to all aspects of indoor environmental quality.

5. Aggressively implement proven sustainable technologies as a matter of course.

Agencies regularly replace worn-out roofs, lighting systems, heating, ventilation, and air conditioning systems, water fixtures, computers, printers, and other equipment in existing buildings. Federal agencies have significant opportunities to upgrade the performance of existing building systems through effective operations, through routine maintenance, repair, and replacement programs, and through retrofit projects. As systems are changed out, more efficient technologies can be incorporated to reduce greenhouse gas emissions and energy and water use, to improve indoor environmental quality, and to meet other objectives related to high-performance green buildings.

6. Use integrated, collaborative processes and practices to overcome conventional segmented processes that fail to optimize resources.

Integrated, collaborative work processes are essential for achieving the multiple objectives associated with high-performance facilities, including zero-net-energy buildings. They can be used to overcome the wasting of resources inherent in conventional, segmented processes and to support a life-cycle perspective.

Agencies could leverage available resources, meet public policy goals, and improve results now and over the long term by consistently implementing existing guidelines such as the "Guiding Principles for Federal Leadership in High-Performance and Sustainable Buildings." Even greater reductions of energy use could be achieved if, during the design process, agencies considered the energy required to operate lighting, computers, servers, copy machines, appliances, and other equipment.

7. Aim for high-performance, near-zero-net-energy buildings now.

The technologies and integrated design processes needed to develop high-performance facilities, including near-zero-net-energy buildings, are already available, and some agencies are using them effectively. Federal agencies that wait until 2020 to begin designing zero-net-energy buildings will be missing a significant opportunity to leapfrog ahead to meet their goals and conserve resources. Starting now also provides the opportunity to learn how best to combine technologies and processes to achieve zero-net-energy buildings for a range of climates and locations, and to share that information with other agencies.

Historic buildings present an opportunity to create zero-net-energy buildings. Many historic structures were originally designed with passive heating and cooling coupled with natural daylighting and ventilation strategies. However, their performance may have been compromised over time through the accretion of mechanical systems and the elimination of original components. By carefully retrofitting and replacing existing systems, some historic structures can become high-performance buildings again.

8. Measure, verify, and report performance to improve processes and change behavior.

Achieving all of the objectives associated with federal high-performance facilities requires changes in mindset as much as it does changes in processes. Change within an organization requires leadership and effective communication so that everyone in the organization understands and accepts that the objectives are the right ones to continuously pursue. Because effective operation of facility systems is dependent, in part, on the behavior of occupants, occupants need to understand how their behavior affects facility performance and why proper operation is important to their own health and safety and to their agency's mission. Best-practice organizations have long used performance measurement as a basis for good communication, for changing conventional processes, and for changing human behavior.

Because an array of performance measures has been developed to track progress toward different goals or objectives related to federal high-performance buildings, some measures conflict and create disincentives for sustainable practices. For example, agencies have been directed to (1) reduce their energy use per square foot of space and (2) reduce their total square footage of space. Reducing total square footage of space should, intuitively, also lead to reduced energy use. However, if an agency is successful in reducing its total square footage of space, its energy use per square foot may increase, and it will appear that the agency is failing to meet the objectives. This lack of alignment among performance measures undermines the achievement of what should be complementary objectives. New performance measures are being developed to track greenhouse gas emissions and carbon footprint. To the extent possible, the government and its agencies should ensure that all performance measures are aligned to achieve complementary objectives.

Other techniques, technologies, and tools that can be used to improve communication and to help change behavior in support of the range of objectives associated with high-performance green buildings are described in Chapter 5.

9. Use performance-based approaches to unleash the creativity of contractors.

When new buildings or major retrofits are needed, federal agencies develop criteria for the projects and then contract with private-sector firms to design and construct them. Federal agencies can use performance-based contracts to set high-level performance goals for new buildings and major retrofits and then challenge private-sector contractors to use their creativity and expertise to design projects that meet those goals.

When several years have elapsed between the actual design of a project and its construction, the designs can become "stale," such that the project will not be state of the art when the "ribbon is cut." In these circumstances, agencies should work with contractors through charrettes or other practices to update the designs to state-of-the-art standards before construction.

10. Collaborate to drive the market for sustainable products and high-performance technologies.

Federal agencies can use their purchasing power to drive the market demand for sustainable products and services. Realizing such opportunities will require agencies to collaborate with each other and with industry, universities, and nonprofits in public-private partnerships.

Agencies can also drive the demand for high-performance space through their leasing practices, as recognized in Executive Order 13514.

Federal agencies have the opportunity to drive the wider deployment of new, more resource-efficient technologies and products by using their facilities as test beds for new technologies and practices and then publicizing the test results. In this way, agencies and the private sector can create a knowledge base for new technologies and practices that will help to mitigate the risk of using them.

11. Use standards and guidelines to drive change and embed sustainability into decision-making processes.

Federal agencies can meet objectives for high-performance facilities by embedding sustainable practices into their policies, design standards, acquisition and maintenance practices, contracts, and task orders, and through the use of guidelines such as green building rating systems.

Many agencies maintain their own sets of design and operations standards to address the types of facilities they typically manage. One relatively easy way to embed sustainability into everyday decision making is to review these standards and revise them as necessary to align with objectives for high-performance green buildings. Specifying Energy Star appliances and equipment, WaterSense fixtures, and FEMP[2]-designated electronics in contracts and task orders would result in improved energy and water performance almost automatically.

12. Communicate successes and learn from others.

Sustainable practices and processes are evolving and proliferating rapidly. Federal agencies have already developed numerous databases and Web sites containing policies, guidelines, processes, tools, technologies, and evidence-based data for developing, operating, retrofitting, and managing high-performance green buildings and facilities. However, these Web sites and databases are scattered among many individual agencies and their overall value is diminished by this segmentation. Federal agencies should collaborate to determine how they can best optimize the value of such information so that it can be used more effectively by all federal agencies and so that it can be easily shared with state and local governments, with private-sector and not-for-profit organizations, and with the public.

---

[2]Federal Energy Management Program.

1

Context

Although environmentally responsive design is a centuries-old concept, the terms "green building," "sustainably designed building," and "high-performance building" have become part of the public dialogue relatively recently. The definitions of these terms vary widely. They can be as detailed as the definition outlined in the Energy Independence and Security Act (EISA) of 2007 (Public Law 110-140) (see Chapter 2) or as broadly as "buildings that are healthy for the occupants, the planet, and for the future of life, and that generate more energy than they use and purify more water than they pollute" (see Appendix D). Overall, the goal is to design buildings that meet a broad range of performance objectives related to land use, transportation, energy and water efficiency, indoor environmental quality, and other factors (NRC, 2007). Advanced efforts such as the Living Building Challenge[1] seek to design buildings that will harvest all of their own energy and water, operate pollution free, and promote the health and well-being of the people who use them.

This new interest in buildings has been spurred, in part, by ever-increasing knowledge about the impact of indoor environments on people and the impacts of buildings on the environment. Research has shown that the design, operation, and maintenance of buildings can affect the health, safety, and productivity of the people who occupy them (NRC, 2007). Leaking roofs, for example, can lead to excessive indoor moisture and mold, which in turn, can exacerbate asthma in people, leading to illness and loss of productivity (NRC, 2007). Buildings are also resource intensive: In the United States, buildings account for 40 percent of primary energy use, 12 percent of water consumption, and 60 percent of all nonindustrial waste (NSTC, 2008). The processes for producing energy at power plants and delivering that energy to buildings to power heating, cooling, ventilation systems, computers, and appliances account for up to 40 percent of U.S. greenhouse gas emissions (NAS-NAE-NRC, 2008). Policy makers and others have recognized that new ways of designing, operating, retrofitting, and managing buildings will be essential in solving the national challenges of energy independence, global climate change, and environmental sustainability.

Greater evidence-based knowledge about buildings has also led to new processes and tools for measuring and evaluating how buildings perform throughout their life cycles. Buildings are systems

---

[1] Information available at http://ilbi.org/lbc.

FIGURE 1.1 Planning for buildings at larger scales; campus of U.S. Department of Energy's National Renewable Energy Laboratory. SOURCE: Pat Corkery.

of interrelated systems and components, including facades, roofs, foundations, windows, mechanical, electrical, ventilation, air conditioning, and plumbing systems. The quality of a building's performance over the 30 or more years it is used will be the result of numerous, individual decisions about location, siting, design, construction, materials, function, operation, and maintenance. Performance will also be a function of how the building is used by occupants.

Today, a building's performance can be measured in terms of its indoor environmental quality (e.g., quality of air, ventilation, lighting, comfort of occupants), its use of materials, energy, and other natural resources, and its emissions into the air and water. Improved evaluation has led to the development of new technologies to reduce greenhouse gas emissions, energy and water use, and to provide power through renewable sources. Among these technologies are "cool" roofs,[2] high-performance lighting, Energy Star rated appliances and equipment, WaterSense fixtures, and windows and control systems that optimize the use of natural daylight while minimizing heat loss.

The scale at which buildings are managed and evaluated is also changing: The focus has shifted from individual buildings to entire portfolios (groups of buildings under a single ownership or management), neighborhoods, communities, regions, watersheds, airsheds, and economies. As the scale of design has increased, so have the opportunities for sharing infrastructure and conserving land and open space. Larger-scale planning allows architects, engineers, planners, and others to leverage infrastructure systems, to cluster development and conserve land and open space, and to think in terms of environmental restoration and regeneration. At larger scales, the use of technologies such as district energy systems, combined heat and power (co-generation) plants, geothermal conditioning systems, water capture and reuse, and others can result in greater reductions of energy and water use than can be realized through a building-by-building approach (Figure 1.1).

FEDERAL FACILITIES

During its 200+ years of existence, the federal government has acquired facilities (buildings and other structures) worldwide to support its various missions and programs for the American public. These facilities enable the conduct of foreign and public policy, national defense, the preservation of historic,

---

[2]Cool roofs include both white roofs, which stay cooler in the sun by reflecting incident sunlight back into space, and green (vegetative) roofs, which absorb rainwater and then cool by evapotranspiration.

cultural, and educational artifacts, scientific and medical research, recreation, and the delivery of goods and services (Figure 1.2). Almost half of these facilities are at least 50 years old and some have been designated historically significant (NRC, 1998).

The U.S. federal government owns or leases approximately 429,000 buildings of many types (Figure 1.3) with a total of 3.34 billion square feet worldwide (Table 1.1), as of fiscal year (FY) 2009. As of that year, approximately 83 percent of the total square footage of federal buildings located within the 50 states was owned space, 13 percent was leased, and 4 percent was otherwise managed (GSA, 2010).

Federal facilities are owned and operated by more than 30 individual departments and agencies. The Army, Air Force, Navy, U.S. General Services Administration (GSA), and Department of Veterans Affairs manage the largest amounts of building space as measured by square footage (GSA, 2010).

Because all federal agencies have different missions and programs, the composition of their individual portfolios of facilities varies widely in terms of building types (e.g., offices, hospitals, barracks, museums, laboratories), age, condition, geographic distribution, and configuration (campus type, installations, individual buildings). The military services, for example, own bases and installations that operate much like small cities and are distributed across the United States and the world. In contrast, the Smithsonian Institution owns museums and research laboratories located in a relatively few, but geographically dispersed locations, while the Department of Energy owns numerous sites containing industrial, administrative, and nuclear facilities. The GSA manages 354 million square feet of space in 8,600 buildings, including offices, border stations, courthouses, laboratories, post offices, and data processing centers. The GSA manages this space on behalf of numerous federal agencies.

Typically, in any given year, the federal government as a whole spends about $30 billion on the design and construction of new facilities (NRC, 2004). In FY 2009, federal agencies reported spending $21.3 billion to operate buildings,[3] of which $13.2 billion was for owned buildings and $8.1 billion for leased buildings[4] (GSA, 2010). However, the level of investment in facilities maintenance and repair has been inadequate for many years, resulting in backlogs of repair projects estimated in the tens of billions of dollars (GAO, 2009). Furthermore, federal agencies reported in FY 2008 that they own more than 10,000 excess buildings (i.e., no longer needed to support agency missions) containing 43 million square feet of space and costing about $133 million to operate. An additional 45,000 buildings have been identified as underutilized (defined as the extent to which a property is used to its fullest capacity) (GSA, 2009).

Recognizing the significant role of buildings in solving national issues such as energy independence and security, and recognizing the opportunity for federal leadership, Congress and two presidential administrations have enacted laws and issued executive orders directing federal agencies to develop high-performance, energy-efficient, and sustainable federal buildings. EISA defines the attributes of a federal high-performance green building and establishes numerous objectives for federal buildings, including objectives for the reduction of energy, water, and fossil fuel use. Executive Order 13514, *Federal Leadership in Environmental, Energy, and Economic Performance*, challenges federal agencies to lead by example to create a clean energy economy. Most recently, a June 2010 presidential memorandum directs federal agencies to accelerate efforts to identify and eliminate excess properties for the purpose of eliminating wasteful spending, saving energy and water, and further reducing greenhouse gas emissions.[5]

---

[3]Operating costs include recurring maintenance and repair costs, utilities, cleaning and/or janitorial costs, and roads and grounds costs (GSA, 2010).

[4]When reporting annual operating costs for leased assets, agencies report the full annual lease costs, including base and operating rent, plus any additional government operating expenses (recurring maintenance and repair costs; utilities; cleaning and/or janitorial costs; roads/ground expenses) not covered in the lease contract.

[5]Available at http://www.whitehouse.gov/the-press-office/presidential-memorandum-disposing-unneeded-federal-real-estate. Accessed February 28, 2011.

FIGURE 1.2 The federal government owns many types of buildings for conducting its missions and programs for the public. From top: Byron Rogers Courthouse, Denver, Colorado; Internal Revenue Service building, Kansas City, Missouri; satellite operations facility, Suitland, Maryland; Library of Congress; U.S. Capitol; White House; Udvar-Hazy Air and Space Museum Annex, Fairfax, Virginia; Arts and Industries Building, Washington, D.C.

TABLE 1.1 Federal Buildings by Predominant Use and
Square Footage as Reported for Fiscal Year 2009

Predominant Use	Square Feet in Millions
Office	740.8
Warehouses	460.4
Service	416.2
Family housing	364.9
Barracks/dormitories	271.2
Schools	251.7
Other institutional uses	221.4
All remaining uses[a]	612.8
Total square feet	3,339.4

[a]All remaining uses include prisons and detention centers, hospitals, laboratories, industrial, communication systems, museums, and post offices.
SOURCE: GSA, 2010.

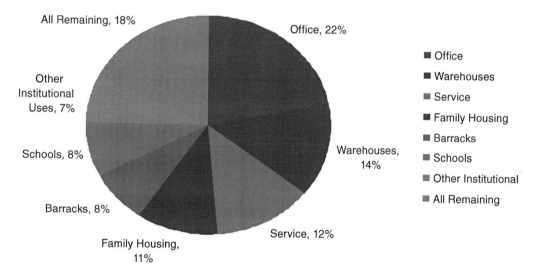

FIGURE 1.3 Federal buildings by predominant use in square feet as reported for fiscal year 2009. SOURCE: GSA, 2010. All remaining uses include prisons and detention centers, hospitals, laboratories, industrial, communication systems, museums, and post offices.

EISA also established the Office of Federal High-Performance Green Buildings within the GSA. As the GSA's green building center of excellence and the federal government's high-performance building thought leader and catalyst, the office "strategically facilitates the adoption of integrated sustainable practices, technologies and behaviors to accelerate the achievement of a zero environmental footprint."[6]

---

[6] From GSA Web site, http://www.gsa.gov/portal/category/101107. Accessed February 28, 2011.

STATEMENT OF TASK

In 2010, GSA's Office of Federal High-Performance Green Buildings asked the National Academies to appoint an ad hoc committee of experts to conduct a public workshop and prepare a report that identifies strategies and approaches for achieving a range of objectives associated with high-performance green federal buildings. To meet its charge, the committee was also asked to identify

- Challenges, barriers, and gaps in knowledge related to developing high-performance green federal buildings.
- Current best practices and ways to optimize resources for achieving high-performance green building objectives during planning, design and construction, and operations and maintenance for new and existing facilities.
- Best practices for reporting the outcomes of investments in high-performance green federal buildings in a transparent manner on public, federal Web sites.
- Approaches, tools, and technologies for overcoming identified challenges, barriers, and gaps in knowledge.

The nine members of the committee have wide-ranging backgrounds in government, industry, and academia and expertise in architecture, engineering, land use planning, facilities program management, construction management, building and energy technologies, and performance measurement (see Appendix A for committee members' biographies).

THE COMMITTEE'S APPROACH

In determining how to fulfill its broad statement of task, the committee recognized that federal agencies have already published other reports and papers, and have created databases and tools related to various aspects of high-performance green buildings. Among these are *Greening Federal Facilities: An Energy, Environmental, and Economic Resource Guide for Federal Facility Managers and Designers* (DOE, 2001), the *Federal Research and Development Agenda for Net-Zero Energy, High-Performance Green Buildings* (NSTC, 2008), the Whole Building Design Guide (www.wbdg.org), the High-Performance Federal Buildings Web site (http://femp.buildinggreen.com/), the Energy Star (www.energystar.gov) and WaterSense (http://www.epa.gov/WaterSense) programs for efficient equipment, appliances, and fixtures, the electronic product environmental assessment tool (http://www.epa.gov/epp/pubs/products/epeat.htm), the Building Energy Software Tools Directory (http://apps1.eere.energy.gov/buildings/tools_directory/), and the newly released Sustainable Facilities Tool (http://www.sftool.org/). In addition, many initiatives are under way within federal agencies and other public- and private-sector organizations, universities, not-for-profit, and community groups, across the country and internationally. To try to capture all of the valuable and thought-provoking ideas, lessons learned, and evidence-based data from these initiatives in three 2-day meetings would not be possible.

The committee determined it would focus on identifying examples of important initiatives taking place and available resources and how these examples and resources could be used to help make sustainability the preferred choice at all levels of decision making. In this way, the report could also be of value to federal agencies with differing missions, types of facilities, and operating procedures: It would be up to the individual agencies to adapt the approaches to their situations.

The committee also decided not to spend significant time researching challenges, barriers, and gaps in knowledge for achieving high-performance green buildings because most are well known and well documented. Nor would it recommend changes to the budget process or other obstacles, which would be outside the scope of this study.

Instead, the committee identified "levers of change," defined as "areas where federal agencies can leverage their resources to spur transformational actions and to make sustainability the preferred choice at all levels of decision making." The levers included systems-based thinking; portfolio-based facilities management; integrated work processes; procurement and finance; communication and feedback for behavioral change; standards and guidelines; and technologies and tools. The levers support an overall life-cycle perspective and can be used in all phases of building design, operation, and management. The levers were chosen, in part, because they can be used immediately by federal agencies to overcome existing challenges and barriers, to achieve objectives related to high-performance buildings, and to support their missions and programs.

A second part of the committee's charge was to "identify current best practices and ways to optimize available resources for achieving high-performance green building objectives." The term "best practice" has been defined differently by different groups and for differing purposes. By some definitions a practice can only be classified as a best practice after it has been used by a variety of organizations over time and has been well documented. Building commissioning, for example, is a well-recognized, well-documented best practice for ensuring that building systems and components are operating as originally designed.

However, new practices, technologies, tools, and processes related to high-performance green buildings are continuously emerging. For this reason, the committee defined best practices more broadly as "processes, procedures, or technologies that optimize available resources and could be effectively applied by the GSA and other federal agencies to meet similar objectives." The committee believed that agencies might lose opportunities to leapfrog ahead to fulfill their mandates if the committee only recommended well-documented best practices with a history of proven results.

At its first meeting, the committee gathered background information on various mandates related to federal high-performance green buildings. The committee also held discussions with staff from the GSA's Office of Federal High-Performance Green Buildings, the Building Technology Research and Development Subcommittee of the National Science and Technology Council, and the Office of the Federal Environmental Executive (see Appendix B for list of committee meetings and speakers).

The committee's second meeting included the public workshop held on the afternoon of July 20, 2010, and on July 21, 2010. The levers of change were used to help structure the agenda. The committee invited speakers representing organizations and groups that had been early adopters of sustainable practices for buildings, installations, and communities. The committee chose the topics of the presentations based on its members' own knowledge and expertise with respect to on-going initiatives; it hoped to show how the levers of change could be implemented. Due to the amount of time available, the number of speakers who could be invited was limited. Many other examples of sustainable initiatives and practices are equally deserving of recognition and study by federal agencies and others.

More than 60 practitioners from public agencies, industry, and academia participated in the workshop and shared their expertise, experiences, and ideas during the breakout sessions (see Appendix C for workshop agenda and participants). The committee also attended the White House Clean Energy Forum on Federal Leaders and Sustainable Building on the morning of July 20, 2010 (information available at http://www.whitehouse.gov/blog/2010/07/21/clean-energy-economy-forum-federal-leaders-and-sustainable-building).

At its third and final meeting in November 2010, the committee heard from several additional speakers representing the Department of State, the Department of Defense, the GSA, the Oberlin, Ohio Project, and the developer of a new tool for life-cycle assessment of supply chains.

Some of the presentations are summarized in Appendixes D through I. The summaries provide more context and detail about specific initiatives and should be read as an integral part of this report.

Chapter 2 focuses on the numerous objectives associated with federal high-performance green buildings and identifies challenges, barriers, and obstacles for achieving those objectives.

2

Objectives and Challenges Associated with Federal High-Performance Green Buildings

OBJECTIVES FOR FEDERAL HIGH-PERFORMANCE GREEN BUILDINGS

The significance of federal buildings in terms of environmental impacts, worker health and productivity, and operating costs has been recognized in an array of federal legislation, executive orders, and guidance documents.

In 2007, the Energy Independence and Security Act (EISA) defined a high-performance green building as one that during its life cycle, as compared with similar buildings (as measured by Commercial Buildings Energy Consumption Survey data from the Energy Information Agency),

(A) reduces energy, water, and material resource use;
(B) improves indoor environmental quality, including reducing indoor pollution, improving thermal comfort, and improving lighting and acoustic environments that affect occupant health and productivity;
(C) reduces negative impacts on the environment throughout the life-cycle of the building, including air and water pollution and waste generation;
(D) increases the use of environmentally preferable products, including bio-based, recycled content, and nontoxic products with lower life-cycle impacts;
(E) increases reuse and recycling opportunities;
(F) integrates systems in the building;
(G) reduces the environmental and energy impacts of transportation through building location and site design that support a full range of transportation choices for users of the building; and
(H) considers indoor and outdoor effects of the building on human health and the environment, including improvements in worker productivity, the life-cycle impacts of building materials and operations, and other factors considered to be appropriate.

Among other provisions, EISA requires that federal agencies reduce their total energy consumption by 30 percent by 2015, relative to 2005 levels. For new federal buildings and major renovations, EISA requires that fossil-fuel energy use—relative to 2003 levels—be reduced 55 percent by 2010 and eliminated altogether (100 percent reduction) by 2030. The EISA standards apply to new construction, major renovations of existing structures, replacement of installed equipment or renovation, rehabilitation, expansion, or remodeling of existing space.

Executive Order 13423, *Strengthening Federal Environmental, Energy, and Transportation Management*,[1] also issued in 2007, requires federal agencies to conduct their environmental, transportation, and energy-related activities under the law in support of their respective missions in an environmentally, economically and fiscally sound, integrated, continuously improving, efficient, and sustainable manner. Among its other provisions, Executive Order 13423 requires federal agencies to

a. improve energy efficiency and reduce greenhouse gas emissions of the agency through reduction of energy intensity by
   i. 3 percent annually through the end of fiscal year (FY) 2015, or
   ii. 30 percent by the end of FY 2015, relative to the baseline of the agency's energy use in fiscal year 2003;

b. ensure that
   i. at least half of the statutorily required renewable energy consumed by the agency in a fiscal year comes from new renewable sources, and
   ii. to the extent feasible, the agency implements renewable energy generation plants on agency property for agency use.

Agencies are also directed to reduce their water intensity (gallons per square foot) by 2 percent each year through FY 2015 for a total of 16 percent reduction below water consumption in FY 2007.

Executive Order 13423 also requires federal agencies to ensure that 15 percent of the existing federal capital asset building inventory of each agency incorporate the sustainable practices outlined in "Guiding Principles for Federal Leadership in High Performance and Sustainable Buildings" (hereinafter called the Guiding Principles) by the end of FY 2015. The Guiding Principles are the following:

(1) Employ Integrated Design Principles;
(2) Optimize Energy Performance;
(3) Protect and Conserve Water;
(4) Enhance Indoor Environmental Quality; and
(5) Reduce Environmental Impact of Materials.[2]

Executive Order 13514, *Federal Leadership in Environmental, Energy, and Economic Performance*, issued in October 2009, challenges federal agencies to lead by example to create a clean energy economy that will increase the nation's prosperity, promote energy security, protect the interests of taxpayers, and safeguard the health of the environment. It states that it is the policy of the United States that federal agencies shall do the following:

- Increase energy efficiency.
- Measure, report, and reduce their greenhouse gas emissions from direct and indirect activities.
- Conserve and protect water resources through efficiency, reuse, and stormwater management.
- Eliminate waste, recycle, and prevent pollution.
- Leverage agency acquisitions to foster markets for sustainable technologies and environmentally preferable materials, products, and services.
- Design, construct, maintain, and operate high-performance sustainable buildings in sustainable locations.
- Strengthen the vitality and livability of the communities in which federal facilities are located.
- Inform federal employees about and involve them in the achievement of these goals.

---

[1]The full text for Executive Order 13423 is available at http://edocket.access.gpo.gov/2007/pdf/07-374.pdf.
[2]The full text for the Guiding Principles is available at http://www.energystar.gov/ia/business/Guiding_Principles.pdf.

Executive Order 13514 also establishes more than 20 facilities-related goals for agencies, among them the following:

- Establishing a percentage reduction target for agency-wide reductions of scope 1[3] and 2[4] greenhouse gas emissions in absolute terms by 2020, relative to a FY 2008 baseline of the agency's scope 1 and scope 2 greenhouse gas emissions.
- Reducing potable water consumption intensity[5] by 2 percent annually through FY 2020, or 26 percent by the end of FY 2020, relative to a baseline of the agency's water consumption in FY 2007, by implementing water management strategies including water-efficient and low-flow fixtures and efficient cooling towers.
- Identifying, promoting, and implementing, consistent with State law, water reuse strategies that reduce potable water consumption.
- Minimizing the generation of waste and pollutants through source reduction.
- Diverting at least 50 percent of construction and demolition materials and debris by the end of FY 2015.
- Participating in regional transportation planning and recognizing existing community transportation infrastructure.
- Ensuring that planning for new federal facilities or new leases includes consideration of sites that are pedestrian friendly, near existing employment centers, and accessible to public transit, and emphasizes existing central cities and, in rural communities, existing or planned town centers.
- Beginning in 2020 and thereafter, ensuring that all new federal buildings that enter the planning process are designed to achieve zero-net-energy by 2030.
- Ensuring that at least 15 percent of the agency's existing buildings (above 5,000 gross square feet) and building leases (above 5,000 gross square feet) meet the Guiding Principles by FY 2015 and that the agency makes annual progress toward 100-percent conformance with the Guiding Principles for its building inventory.
- Pursuing cost-effective, innovative strategies, such as highly reflective and vegetated roofs, to minimize consumption of energy, water, and materials.
- Identifying opportunities to consolidate and dispose of existing assets, optimize the performance of the agency's real-property portfolio, and reduce associated environmental impacts, when adding assets to the agency's real property inventory.
- Ensuring that rehabilitation of federally-owned historic buildings utilizes best practices and technologies in retrofitting to promote long-term viability of the buildings.
- Advancing sustainable acquisition to ensure that 95 percent of new contract actions including task and delivery orders, for products and services with the exception of acquisition of weapon systems, are energy-efficient (Energy Star or Federal Energy Management Program (FEMP) designated), water-efficient, bio-based, environmentally preferable (e.g., Electronic Product Environmental Assessment Tool (EPEAT) certified), non-ozone depleting, contain recycled content, or are non-toxic or less-toxic alternatives, where such products and services meet agency performance requirements.
- Ensuring the procurement of Energy Star and FEMP-designated electronic equipment.

The executive order requires each agency to develop, implement, and annually update an integrated Strategic Sustainability Performance Plan that prioritizes agency actions based on life-cycle return on investment. The plans are to be integrated into the agency's strategic plan and to meet additional requirements, as specified.[6]

The June 2010 Presidential memorandum "Disposing of Unneeded Federal Real Estate—Increasing Sales Proceeds, Cutting Operating Costs, and Improving Energy Efficiency"[7] directs federal agencies to accelerate efforts to identify and eliminate excess properties in order to eliminate wasteful spending of taxpayer dollars, save energy and water, and further reduce greenhouse gas pollution. It also estab-

---

[3]Scope 1 emissions are defined as direct greenhouse gas emissions from sources that are owned or controlled by the federal agency.

[4]Scope 2 emissions are defined as direct greenhouse gas emissions resulting from the generation of electricity, heat, or steam purchased by a federal agency.

[5]Water consumption intensity is defined as water consumption per square foot of building space.

[6]A Crosswalk of Sustainability Goals and Targets is available at http://www1.eere.energy.gov/femp/pdfs/sustainabilitycrosswalk.pdf.

[7]Available at http://www.fedcenter.gov/admin/itemattachment.cfm?attachmentid=307.

lishes a target of saving $3 billion government-wide through the disposition of excess buildings, space consolidation, and other methods by the end of FY 2012.

Most of these mandates have not been funded. The exception is the American Reinvestment and Recovery Act of 2009 which directed $5.5 billion to the General Services Administration to convert federal facilities into high-performance green buildings and to create jobs in the architecture-engineering-construction industry. Additional billions of dollars were allocated to the Department of Defense and other federal organizations to improve the energy efficiency and other characteristics of their facilities. All indications are that future funding for many federal agencies and their programs will decrease.

CHALLENGES AND BARRIERS

The development of high-performance federal facilities has the potential to result in substantial long-term cost savings and cost avoidances through more efficient use of energy, water, and other resources. Before this can happen, however, federal agencies will need to find ways to overcome a range of challenges and barriers. Most of these are long-standing and well documented. All are interrelated.

Embedding Sustainability into Everyday Decision Making

In the federal government, those who directly influence federal facilities investments include department and agency senior executives, facilities program managers, budgeting and financial analysts, Congress, the President, other policy makers, and special interest constituencies (NRC, 2004). In this decision-making structure, the various government entities have diverse but overlapping objectives. As shown in Figure 2.1, some decision-making and operating groups, such as the Office of Management and Budget (OMB) and the Congressional Budget Office (CBO), focus on balancing the budget, while departments and agencies focus on issues related to their missions.

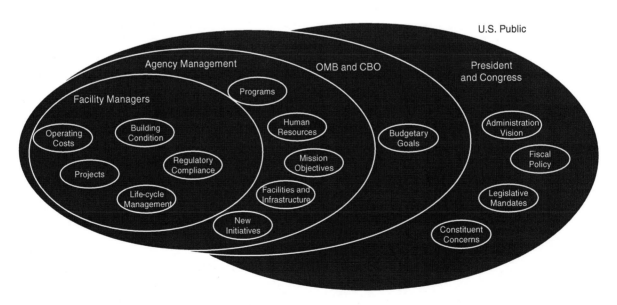

FIGURE 2.1 The various stakeholders in federal facilities investments and their diverse and overlapping objectives. SOURCE: NRC, 2004.

A primary challenge for the federal government and its agencies is to find ways to embed sustainable thinking into all processes such that sustainability becomes the preferred choice of decision makers at all levels, across a number of organizations with different objectives.

Excess Facilities That Siphon Off Resources

The fact that federal agencies own more facilities than required to support their current and future missions is a long-standing and well-documented issue (NRC, 1998, 2004; GSA, 2010; GAO, 2011), most recently recognized in the June 2010 Presidential memorandum "Disposing of Unneeded Federal Real Estate."

This issue takes on a new urgency in an era of constrained resources: Unneeded facilities use energy and water, materials, and staff time, and siphon off funding that could be better used to retrofit buildings that directly support agencies' missions and programs.

That agencies have found it difficult to dispose of unneeded facilities is well documented. The reasons for this include myriad regulations for transferring title to nonfederal entities, disincentives created by the budget structure, security issues related to the location of some excess facilities, and the condition of some facilities (NRC, 1998). Additional issues include the "numerous stakeholders that have an interest in how the federal government carries out its real property acquisition, management, and disposal practices" and a "complex legal environment that has a significant impact on real property decision making and may not lead to economically rational outcomes" (GAO, 2011, p. 5).

The most visible and far-reaching effort to dispose of unneeded federal facilities has been the Base Realignment and Closure (BRAC) process. Through BRAC, the Department of Defense, in conjunction with an independent commission, first determined which facilities were needed to support current and future missions (alignment) and then identified which facilities were excess and should be closed. This process was not without controversy. Any future activities to dispose of excess federal facilities on a large scale will also raise many issues and concerns on the part of the public, Congress, federal agencies, and other stakeholders.

Federal Budget Process

Each of the mandates related to federal high-performance green buildings specifically calls for the use of a life-cycle perspective or life-cycle costing. A life-cycle perspective involves consideration of all phases of a building's life cycle: programming/planning, design, construction, operations, maintenance and repair, retrofit, and demolition or deconstruction. Life-cycle costing for buildings involves looking at the interactions, costs, and performance of all of its components, from planning through operations, through disposal.

The phases in a building's life cycle vary widely in time, costs, and use of resources. So, while it may take 1 to 7 years to plan, design, and construct a building, once built it will be used for 30 years or longer. During the time a building is used, the costs to operate, maintain, and repair it will be six to eight times greater than the initial cost of design and construction (NRC, 2004). For that reason alone, a focus on the life-cycle costs of buildings, not just the first costs of design and construction, is important for effective decision making and the long-term economic health of the organizations that own them.

One significant barrier to effective implementation of a life-cycle perspective and for life-cycle costing is the federal budget process, which is structured to focus only on the first costs (design and construction) of new buildings and major retrofits. During the budget process, agencies' funding requests for the design and construction of new buildings and major retrofits are considered on a case-by-case basis under separate line items. In contrast, funding requests to operate, maintain, repair, or demolish facilities are lumped together in a different line item that collectively applies to all existing facilities. As a result, up to 85 percent of the total life-cycle costs of buildings are not transparent to or routinely considered by executive decision makers (NRC, 2004).

The focus on first costs is reinforced by the budget "scorekeeping" procedures mandated as part of the Budget Enforcement Act of 1990. Scorekeeping is a process for estimating the budgetary effects of pending and enacted legislation. Scoring a proposal for funding a new facility or major retrofit is intended to provide the transparency needed for effective congressional and public oversight. In fact, the scoring of major facilities proposals discloses only the projected design and construction costs, not the full life-cycle costs (NRC, 2004).

Scorekeeping procedures also hinder energy supply and technology provisions in funding authorization bills (PCAST, 2010). A recent report of the President's Council of Advisors on Science and Technology (PCAST) recommended that OMB "should develop criteria for determining the life-cycle costs and for including social costs in evaluating energy purchases, and should incorporate this methodology into agency procurements so that the federal government maximizes its influence on clean energy development that is most economical in the long run" (PCAST, 2010, p. 20).

Scorekeeping procedures create incentives for agencies to drive down the first costs of facilities, even if doing so drives up operation and maintenance costs, in order to lessen the impact on the current-year budget (NRC, 2004). In this way, the scorekeeping procedures can indirectly increase the long-term costs of facilities operations and maintenance.

Another budget-related challenge is the lag between the time when an agency identifies the need for a new building or major retrofit and the time when funding is received to construct it. In the federal government, this time period can last as long as 5 to 7 years (Figure 2.2). In the interim, new approaches, technologies, and evidence-based knowledge for high-performance buildings can be developed, and

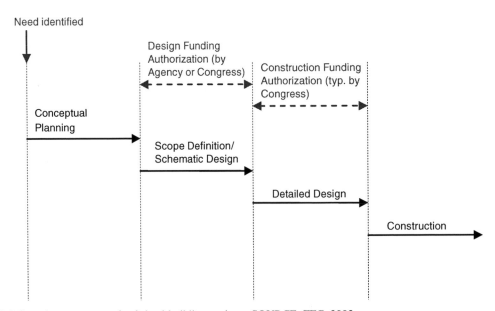

FIGURE 2.2 Generic process map for federal building projects. SOURCE: FFC, 2003.

project designs can become stale by sitting on a shelf. If an agency proceeds with construction but fails to update the design, a newly constructed or renovated building can be obsolete at initial occupancy.

The time lag between funding for design and funding for construction also makes it difficult for agencies to use project delivery processes such as design-build, which undertake design and construction on concurrent paths, not as separate processes. Because funding for design and construction is often allocated in different years, agencies use bridging documents or other work-arounds that are inherently less efficient than design-build is intended to be.

Segmented Processes That Fail to Optimize Resources

Segmentation in all phases of building design and operation is pervasive, long-standing, and well documented (NRC, 2009). Today, most buildings are created and retrofitted through a series of phased, segmented processes: programming/funding, design, construction, operations, maintenance, retrofit, and disposal or demolition. Each phase involves different actors with different objectives, different processes, and differing incentives (Figure 2.3).

This level of segmentation fails to optimize the resources invested in buildings—time, materials, staff expertise, technologies, funding—and also results in greater room for error, lost opportunities for innovation, and less than optimal building performance.

Current federal practices treat the decision making and funding for building projects separately

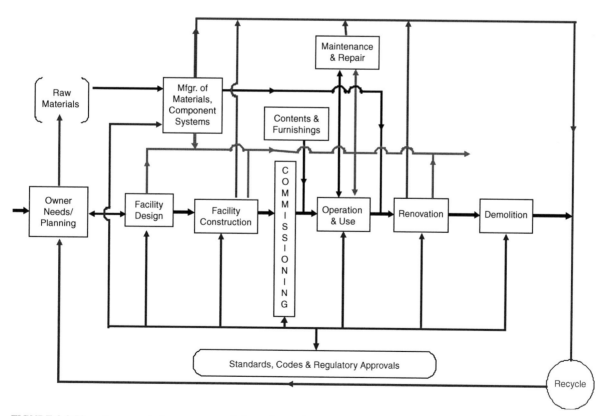

FIGURE 2.3 Phased, segmented processes used throughout a building's life cycle. SOURCE: Originally adapted from a presentation by NIST. Published in NRC, 2009.

from those for building furnishings, equipment, and operations. This is significant because over the life of a facility, computers, copy machines, lighting, and other equipment will use substantial amounts of energy, water, and other resources to operate. The fastest growing segment of energy use in buildings is, in fact, the energy used to power electronics, computers, and servers (DOE, 2008).

Because the energy used by computers and servers, electronics, equipment, lighting, and appliances typically are not accounted for up front in the planning process, agencies are losing a significant opportunity to reduce their total energy use and to meet their mandates.

Lack of Alignment Between Reporting Requirements and Performance Measurement Systems

Each mandate related to federal high-performance buildings sets goals and targets for meeting those goals, and it establishes baselines and performance indicators for measuring progress toward the goals. Although the overall intent of these mandates may be similar, the accretion of these mandates over time has resulted in several different obstacles for creating high-performance buildings.

One obstacle is the multitude of reporting requirements and different baselines. Agencies are required to track and report total energy consumption, greenhouse gas emissions, total water consumption, reduction of fossil fuel use, use of renewable energy, and so forth. To measure progress in meeting a variety of goals, agencies must use a variety of baselines: EISA, for example, requires agencies to measure the reduction of energy and fossil fuel use against a FY 2003 baseline. Executive Order 13423 requires federal agencies to reduce their total water use and to track progress against a FY 2007 baseline, and Executive Order 13514 sets a baseline year of FY 2008 for tracking greenhouse gas emissions. A good deal of staff time is spent in creating the various baselines and tracking and reporting the results. Unfortunately, effective, transparent communication with decision makers and the public about what is actually happening becomes problematic when a variety of resources and baselines are involved.

The accretion of policies, guidelines, standards, and legislation over time has also led to a lack of alignment among performance measures. In some cases, the measures are in conflict or create disincentives for sustainable practices. For example, agencies have been directed to (1) reduce their energy use and (2) reduce the number of square feet they occupy. Reducing total square footage should, intuitively, reduce energy use. However, progress in reducing energy use is measured in terms of energy used per square foot. Even if an agency is successful in reducing its total square footage, its energy use per square foot may increase and it will appear that energy use is actually going up. This lack of alignment among performance measures undermines the achievement of what should be complementary objectives: reducing energy consumption and reducing total building square footage.

Some mandates establish interim goals with the intent of ensuring that federal agencies are making progress in meeting more ambitious, long-term goals. If agencies make significant capital investments focused only on interim goals, they risk long-term sub-optimization: Once major systems or components have been incorporated, it is difficult and expensive to change them again in less than 20 years.

Perceived Higher Costs of Building Green

A significant barrier to developing high-performance green buildings is the widespread perception that green buildings cost much more to design and construct than conventional buildings. A 2010 study documented that people thought green buildings cost 17 percent, on average, more up front. However, the same report gathered evidence-based data for 146 green buildings which showed that the actual cost premium was closer to 2 percent of total design and construction costs (Figure 2.4). Over the lifetime of

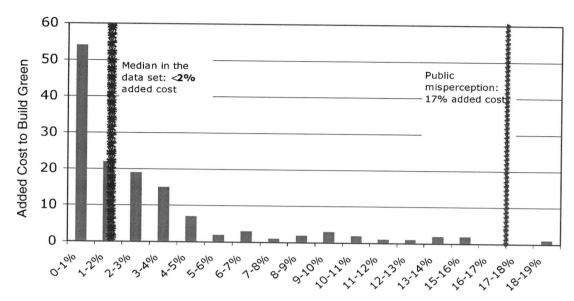

FIGURE 2.4 Cost of building green: evidence from 146 green buildings. SOURCE: Kats, 2010, and Appendix I.

a green building, the savings in energy use alone will far outweigh the initial 2 percent premium (Kats, 2010; Appendix I of this report).

The overall lack of evidence-based data to support the case for a premium up front in order to achieve life-cycle cost savings has made it difficult for federal agency managers to make a business case for high-performance buildings. Similarly, the cost of using renewable energy sources can be higher than the cost of fossil fuels, making it equally difficult for agencies to make a business case based solely on financial return on investment.

Workforce Skills and Occupant Behavior

Effective use of new technologies and new processes requires a workforce that is adequately trained to make decisions and implement them to maximum benefit. Facilities managers must also be adequately trained to operate systems at their optimal level of performance and to understand how things work so that they can fix problems. At the workshop, federal agency managers reported that they were unable to hire enough resource managers who understand new technologies or the interaction of complex building systems well enough to operate them effectively. One report (NRC, 2008) recognized that the skills of the federal workforce are not well aligned with the requirements of 21st century facilities asset management and that additional training is needed (NRC, 2008).

The effective operation of high-performance facilities is also dependent, in part, on building occupants. Occupants can easily undermine effective building operations by bringing in additional appliances and equipment (e.g., heaters, fans, coffee pots), by leaving computers and lights on, and similar practices. Occupants need to understand how their behavior can affect effective facility operations and, in turn, how facility performance can affect their health and productivity.

Widespread Deployment of Innovative Technologies for High-Performance Buildings

Meeting mandates to reduce energy and water use in facilities, to reduce greenhouse gas emissions, and to use environmentally preferable products requires the use of new technologies and products. The barriers to deploying innovative technologies on a widespread basis are well known and well documented. The market for technologies for high-performance facilities consists of many different components designed to achieve different purposes, thousands of developers and suppliers, and potentially hundreds of millions of users (NAS-NAE-NRC, 2010). The size and complexity of this market, the lack of knowledge about the effects of some of these technologies, and risk-averse behavior on the part of suppliers and purchasers limits the adoption of new technologies and tools. Other factors that come into play include the following:

- Limited supply and availability of some technologies;
- Consumers' lack of information and lack of time to do the necessary research;
- Lack of capital for investment;
- Fiscal or regulatory policies that discourage investment in high-performance technologies, even inadvertently;
- Building codes and standards; and
- Perceived risk of new technologies and concerns about legal claims and liability in the event of failure (NAS-NAE-NRC, 2010).

Gaps in Knowledge

Although much progress has been achieved in all facets of creating high-performance facilities, additional research is needed about processes, metrics, and evidence-based design, along with additional testing and development of new tools and technologies. The *Federal Research and Development Agenda for Net-Zero Energy, High-Performance Green Buildings* (NSTC, 2008) addresses a range of research and development needs related to

- Effective performance measures and metrics;
- Net-zero-energy building technologies and strategies;
- A scientific and technical basis for significant reductions in water use and improved rainwater retention;
- Processes, protocols, and products for building materials that minimize resource utilization, waste, and life-cycle environmental impacts;
- A knowledge base and associated energy efficiency technologies and practices needed to promote occupant health, comfort, and productivity; and
- Technology transfer.

Additional areas require more research and development. For example, the interdependencies among systems are generally unknown, which creates uncertainty and reduces willingness to invest in the commercialization of promising technologies.

Chapter 3 discusses the levers of change that can be used by federal agencies to overcome the identified challenges, barriers, and gaps in knowledge so that they can meet objectives related to high-performance facilities.

3

Levers of Change

The committee was asked to identify "approaches, tools, and technologies for overcoming identified challenges, barriers, and gaps in knowledge." To do so, the committee devised levers of change, which it defined as "areas where federal agencies can leverage their resources to spur transformational actions and to make sustainability the preferred choice at all levels of decision making." As noted in Chapter 1, the intent was not to recommend changes to the budget process or to directly confront other challenges, which was outside the scope of the study, but to find ways for federal agencies to overcome such barriers and achieve a range of objectives related to high-performance green buildings. The committee's levers of change include

- Systems-based thinking,
- Portfolio-based facilities management,
- Integrated work processes,
- Procurement, contracting, and finance,
- Communication and feedback for behavioral change,
- Standards and guidelines, and
- Technologies and tools.

Although technologies are themselves a lever of change, they are best enabled through the other levers. All the levers are discussed below.

SYSTEMS-BASED THINKING

Systems-based, or holistic, thinking is aimed at bringing coherence and integration to an area of study to develop a better understanding of its nature and function. By focusing on an entire system, its components, and its ramifications, it becomes possible to look at how efficiently the system uses resources (financial, people, technology, materials, energy), to eliminate waste, and to manage environmental impacts. Effective systems-based thinking begins with the development of goals and objectives for the activity: The more ambitious the goals, the more innovative the solutions are likely to be. Executive Order 13514 implies the use of systems-based thinking by directing federal agencies to develop a comprehensive inventory of greenhouse gas emissions associated with their supply chains.

Systems-based thinking provides a life-cycle perspective that can overcome challenges posed by the budget process and by segmented work processes. It can help federal agencies meet ambitious mandates for the environment and quality of life by providing a more comprehensive understanding of the use of resources and their interrelationships. This understanding, in turn, can help agencies identify new ways to use resources, to substitute more sustainable resources, and to reduce their total use. In the process, agencies can find innovative solutions that will meet a variety of objectives as opposed to finding narrowly focused solutions with unintended consequences.

The difference between conventional thinking and systems-based thinking is apparent when determining how to reduce energy use. In conventional thinking, the use of electricity or natural gas is typically measured by meters at the point of delivery, and total energy for heating (gas for a furnace, electricity for furnace fans or hot water pumps), cooling (typically all electric), lighting (all electric), and appliances and computers is measured. Efforts to reduce energy use at the site typically focus on reducing the energy use per square foot of floor space and do not consider the source of energy. Energy savings are achieved by using equipment and appliances that are energy efficient—for example, Energy Star appliances and equipment and Federal Energy Management Program (FEMP)-designated electronic products.

In systems-based thinking, the focus is on the source of the energy and how efficiently resources are used to produce and deliver energy to a building (Figure 3.1).

For example, to produce electricity, coal is typically burned in a power plant to generate heat and to produce steam. The steam is then turned into mechanical energy to operate a turbine that generates electricity. In this process, about 65 percent of the original energy is typically lost in the form of waste heat emitted through smokestacks and cooling towers. As electricity moves along transmission lines to arrive at buildings, additional energy losses occur. As shown in Figure 3.1, by the time the electricity lights an incandescent bulb, the light produced represents less than 2 percent of the energy used to produce it (NAS-NAE-NRC, 2008). In contrast, the direct delivery of natural gas to a building to produce light would be more efficient and less wasteful.

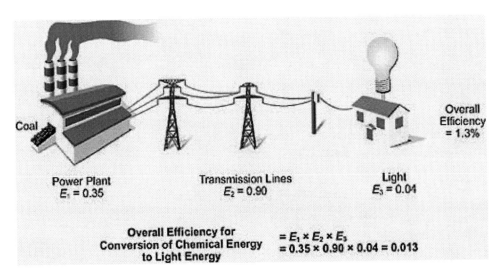

FIGURE 3.1 Applying systems-based thinking to the use of electricity. SOURCE: NAS-NAE-NRC, 2008.

Figure 3.1 illustrates why there are two conventions—primary (or "source") and delivered (or "site")—for attributing energy to each kilowatt hour (kWh) of electricity delivered to the customer's meter. (In the United States this energy is still usually measured in British thermal units, or Btu.) Figure 3.1 assumes that enough coal is burned to provide 1 Btu of heat to a power plant with an efficiency of 35 percent so the energy delivered to the transmission line is 0.35 Btu. This 0.35 Btu is then transmitted and delivered to the meter with an efficiency of 90 percent, for an overall system efficiency of $E1 \times E2 = 31.5$ percent, which is often rounded off to one-third. Thus, 3 Btu of coal burned at the power plant deliver only 1 Btu to the customer's site, and waste 2 Btu in cooling towers and hot stack gasses.

For federal agencies and other organizations, the question to be answered is Which measure should be used when the goal is to minimize energy intensity per square foot of floor area? If a building is all electric this would not matter, but most buildings use both electricity and natural gas. The owner or customer pays not only for the delivered kWh but also for the wasted energy. From a systems point of view it makes better sense to use the primary fuel metric when setting energy-saving goals.

Figure 3.2 illustrates how the priority of a project depends on which metric is used. According to the site energy use line (red), heating, which is mainly by natural gas, uses more energy per square foot than lighting, which is all electric. But the primary energy use line (blue) shows that it is lighting which is the most energy intensive. Systems-based thinking would prioritize attention to electric lighting to reduce costs and greenhouse gas emissions.

Systems-based thinking can also be applied to water use. In conventional thinking, potable water, which is chemically processed at treatment plants and transported to building sites, is used not only for drinking but also for building equipment and fixtures and to irrigate landscaping. A systems-based approach, in contrast, considers using rain- and stormwater for purposes other than drinking. In a systems-based approach, the goal is to use potable water at least twice (first for drinking, then for filtered

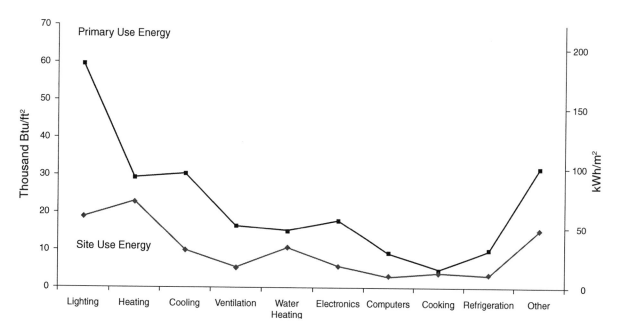

FIGURE 3.2 U.S. commercial buildings' energy consumption by end use in 2006. NOTE: For end uses that are entirely electric the blue point (consumption of primary energy) is simply three times higher than the red point. For uses like cooking with natural gas, the points coincide. For mixed gas/electric uses the points approach one another. SOURCE: DOE, 2008.

grey water uses[1]). Filtered grey water can be used in building equipment and for landscape irrigation. In this way, a systems-based approach can reduce the demand for potable water and, in turn, reduce the energy costs and the cost of chemicals involved in processing and transporting potable water to a site. Managing all stormwater on site can reduce flooding due to stormwater runoff from building sites. Less flooding, in turn, contributes to reducing combined storm-sewer overflow. Systems-based thinking can be applied to many other processes, including waste management and transportation, and it can be applied at many levels, from global and regional to communities, buildings, and supply chains.

A corollary to systems-based thinking for innovation in the built environment is life-cycle accounting. Agencies could pursue three levels of life-cycle accounting to help meet their mandates related to high-performance buildings:

- Traditional life-cycle costing with operational costs factored into first-cost trade-offs;
- Life-cycle assessment that evaluates a full range of environmental consequences from global warming and ozone depletion to habitat reduction and human health; and
- Triple bottom line accounting in which net present value calculations are completed three times, weighing hard economic benefits for the owner first, softer environmental benefits for society second, and known human benefits such as health, productivity, and even jobs last of all.

PORTFOLIO-BASED FACILITIES MANAGEMENT

Just as the scope and scale of building design are evolving, so is the focus of facilities management. Much of this change has occurred in parallel with the growth of information technology and with the increased expectations of facility owners and users for building performance and cost effectiveness (NRC, 2008). In the last 20 years, public and private organizations with large inventories of facilities have shifted from managing individual buildings to managing entire portfolios of facilities (Figure 3.3).

The shift to portfolio-based management has been driven, in part, by the recognition of the costs of facilities, the role of facilities in organizational operations, and the impacts of facilities on workforce health and safety. This recognition has led to a more strategic approach that views facilities as assets that enable the production and delivery of goods and services. Portfolio-based facilities management has been defined as a

> Systematic process of maintaining, upgrading, and operating physical assets cost effectively. It combines engineering principles with sound business practices and economic theory, and provides tools to facilitate a more organized, logical approach to decision making. A facilities asset management approach allows for both program or network-level management and project-level management and thereby supports both executive-level and field-level decision making. (NRC, 2004, p. 32)

Effective portfolio-based facilities management looks holistically at the entire inventory of existing buildings and considers new investments within this context. Life-cycle costing is used for all potential investments. Portfolio-based facilities management can be used to align facilities with missions, to identify excess facilities and underutilized space, to limit the construction of new space, and to identify opportunities for consolidating space.

Well-designed facilities portfolio management programs start with a clear framing of facilities-management goals linked to overarching organizational goals and missions. The goals are used as a

---

[1]Grey water is wastewater from hand washing, showers, and kitchen appliances such as dishwashers and washing machines. It does not include water from toilets.

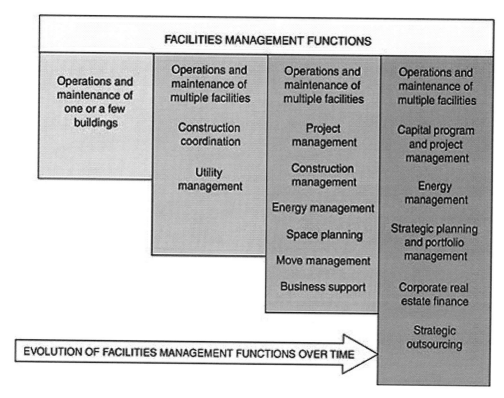

FIGURE 3.3 Evolution of facilities management functions. SOURCE: NRC, 2008.

basis for decision making through all aspects of facilities management, from planning and programming, design and construction, operations, maintenance and repair, retrofit, and demolition.

During planning and programming for a new activity, an agency using a portfolio-based approach first determines if the new activity or program can be accommodated within the existing portfolio of buildings or if the activity can be provided through alternative noncapital solutions such as operations scheduling or leasing, or by using Web-based technologies. By not constructing a new building, agencies can realize multiple benefits including the avoidance of a building's life-cycle costs, and environmental impacts.

When new facilities are needed, the choice of location and site for the facility will have implications for the ultimate sustainability of the building and its total life-cycle costs. For example, the local climate will determine the types and amounts of natural resources that can be drawn upon for the building design (e.g., amount of sunshine, local temperatures), and locations near public transportation may reduce the space needed for on-site parking. The size of the site will also help determine how effectively natural resources, such as daylight, wind, and water, can be used to reduce energy and water use in one or more buildings.

The importance of site selection for building new high-performance facilities is recognized in the Energy Independence and Security Act of 2007, the "Guiding Principles for Federal Leadership in High Performance and Sustainable Buildings," and in Executive Order 13514. In April 2010, "Recommendations on Sustainable Siting for Federal Facilities" was published. The document was a collaborative effort of the U.S. Departments of Transportation, Housing and Urban Development, Defense, and Homeland Security, the General Services Administration (GSA), and the Environmental Protection Agency. It is

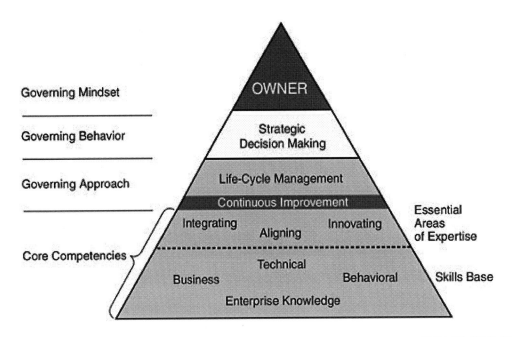

FIGURE 3.4 Recommended framework for effective federal facilities asset management. SOURCE: NRC, 2008.

intended to fulfill Section 10 in Executive Order 13514 which calls for providing the chair of the Council on Environmental Quality with recommendations for sustainable location strategies for consideration in federal agency sustainability plans.[2]

An NRC report (2008) found that to fully implement a facilities portfolio asset management approach, federal agencies require a workforce with a set of core competencies in three areas of expertise and with a skills base. The three areas of expertise are

- Integrating people, processes, places, and technologies by using a life-cycle approach;
- Aligning the facilities portfolio with the organization's missions and available resources; and
- Innovating across traditional functional lines and processes to address changing requirements and opportunities.

The skills base includes a balance of technical, business, and behavioral capabilities along with enterprise knowledge. Enterprise knowledge includes an understanding of the facilities portfolio and how to align it with the organization's mission; of the organization's culture, policy framework and financial constraints; of agency inter- and intra-dependencies; and of the workforce's capabilities and skills (NRC, 2008) (Figure 3.4).

The Federal Buildings Personnel Training Act of 2010 (Public Law 111-308)[3] directs the GSA, in consultation with others, to identify the core competencies necessary for federal personnel performing building operations and maintenance, energy management, safety, and design functions. The competencies include those related to building operations and maintenance, energy management, sustainability, water efficiency, safety (including electrical safety) and building performance measures. The Act also

---

[2]The document is available at http://www.dot.gov/livability/docs/siting_recs.pdf.
[3]The full text of the law is available at http://www.gpo.gov/fdsys/pkg/PLAW-111publ308/pdf/PLAW-111publ308.pdf.

specifies that not later than 18 months after the date of enactment, and annually thereafter, the GSA in conjunction with the Department of Energy, and in consultation with others, shall develop a recommended curriculum for facility management and the operation of high-performance buildings.

INTEGRATED WORK PROCESSES

Integrated work processes are essential for achieving the multiple objectives associated with high-performance buildings. They can be used to overcome the failure to optimize resources inherent in conventional, segmented processes; to support a life-cycle perspective; and to overcome time lags created by the budget process. A recent report of the National Academies found that

> The main difference between high-performing buildings and conventional buildings is essentially an attention to integration, interaction, and quality control throughout the design, construction, and operation of a building. This process, typically referred to as *integrated design*, represents a transformation not in technology but in conceptual thinking about how building systems can most effectively work together and the successful implementation of design intent. (NAS, 2010, p. 96)

The value of integrated processes has been widely recognized. For example, the "Guiding Principles for Federal Leadership in High Performance and Sustainable Buildings" directs federal agencies to use a collaborative, integrated planning and design process that

- Initiates and maintains an integrated project team in all stages of a project's planning and delivery;
- Establishes performance goals for siting, energy, water, materials, and indoor environmental quality along with other comprehensive design goals;
- Ensures incorporation of these goals throughout the design and life cycle of the building; and
- Considers all stages of the building's life cycle, including deconstruction.

The American Institute of Architects (AIA) defines integrated project delivery as

> a project delivery approach that integrates people, systems, business structures, and practices into a process that collaboratively harnesses the talents and insights of all participants to reduce waste and optimize efficiency through all phases of design, fabrication, and construction. (AIA, 2007, p. 1)

The International Council for Research and Innovation in Building and Construction (CIB)[4] recently launched the theme "Improving construction and use through integrated design solutions." The CIB stated that

> Integrated design solutions use collaborative work processes and enhanced skills, with integrated data, information, and knowledge management to minimize structural and process inefficiencies and to enhance the value delivered during design, build, and operation, and across projects. (CIB, 2009, p. 1)[5]

Collaboration is a common theme through all of the above definitions. To be effective, an integrated design process uses a design team having diverse expertise and perspectives: the owner's representatives, contractors, architects, engineers, land use planners, interior designers, facilities managers, preservationists, procurement, finance, and security specialists, among others. Such a team must be able

---

[4]More information available at http://www.cibworld.nl/site/home/index.html.

[5]Available at http://heyblom.websites.xs4all.nl/website/newsletter/0907/ids2009.pdf.

to work collaboratively to achieve a given set of crosscutting and interrelated objectives. In addition, they must understand the interactions of building systems and technologies, which technologies will have immediate paybacks and which will have longer range paybacks in order to make trade-offs and well-informed decisions.

Integrated design processes can be used with any project delivery method (e.g., design-build, design-bid-build). Such efforts require more up-front planning and time commitment than conventional processes. However, the benefits can be immediate and long-lasting. A study published by the National Science and Technology Council (NSTC) found that

> There is a limit to the overall energy savings potential of mainstream approaches for reducing energy use in new buildings. Major national studies agree that this limit ranges from 30% to 50% . . . . Integrating technologies with the building design (form) to create a building that delivers efficiency as a single system, however, can raise savings to 70% of building energy use compared with conventional new design. (NSTC, 2008, p. 17)

PROCUREMENT, CONTRACTING, AND FINANCE

Federal agencies spend billions of dollars annually to procure furnishings, equipment, computers, and other facility-related products. They also spend billions of dollars to lease physical space and to contract with private-sector firms for design, construction, operations, and maintenance services.

The use of procurement and contracting methods as a lever of change is recognized in Executive Order 13514, which specifically directs agencies to leverage their acquisitions to foster markets for sustainable products and to ensure that new contract actions advance energy-efficient, water-efficient, and environmentally preferable products. Specifying Energy Star appliances and equipment, WaterSense fixtures,[6] and FEMP-designated electronics in contracts and task orders would result in improved energy and water performance almost automatically.

The Executive Order also ties the leasing of physical space to achievement of the Guiding Principles.

COMMUNICATION AND FEEDBACK FOR BEHAVIORAL CHANGE

Implementing systems-based thinking, portfolio-based facilities management, and integrated work processes requires changes in mindset throughout federal agencies. Change within an organization requires leadership and effective communication so that all members of the organization understand and accept that the goals and objectives are the right ones to pursue. Because a facility's overall performance is, in part, a function of how the occupants use it, occupants need to understand how proper operation of facility systems (e.g., HVAC systems and controls) can affect their health and productivity, and how they can help achieve the goals for high-performance facilities. If sustainable practices are to become embedded in decision making at all levels of government, facilities managers and other federal staff will need to be more effective in communicating how high-performance facilities enable the agency's mission.

Best-practice organizations have long used performance measurement as a basis for good communication, for changing conventional processes, and for changing human behavior. Performance measures help to identify where objectives are not being met or where they are being exceeded. Managers can then investigate the factors or reasons underlying the performance and make appropriate adjustments. Continuous process monitoring and feedback is also necessary because however effectively one plans, unintended consequences, unforeseen events, and change will occur (NRC, 2004).

---

[6]Additional information available at http://www.epa.gov/WaterSense/.

STANDARDS AND GUIDELINES

Federal agencies can embed sustainability into day-to-day decision making through the use of standards and guidelines. For example, many federal agencies maintain their own sets of design and operations standards to address the types of facilities that they typically manage or contract for. Over time, the standards are updated to embed lessons learned from the design and operation of buildings in order to replicate successes and eliminate failures. Because of the relative newness of some sustainable design and operations practices, existing standards may not be supportive of some aspects of high-performance facilities and associated technologies. One relatively easy way to make sustainability the preferred choice is to review existing standards and revise them as necessary to ensure they support the development of high-performance facilities.

More than 10 rating systems for green buildings that offer certifications for building performance have been developed worldwide (IFMA Foundation, 2010). Three systems available for buildings outside their home countries are the Building Research Establishment Environmental Assessment Method (BREEAM, www.breeam.org), Leadership in Environmental and Energy Design (LEED, www.usgbc.org), and Green Globes (www.greenglobes.com) (IFMA Foundation, 2010).

A number of federal agencies are using the LEED or Green Globes guidelines. Some have established policies that require new buildings or major retrofits to achieve a particular level of LEED—for example, Certified, Silver, Gold, or Platinum. Some require third-party certification of the design, while others self-certify.[7]

Two new standards are available for use by federal agencies. The American National Standards Institute and the American Society of Heating, Refrigerating, and Air-Conditioning Engineers have issued Standard 189.1-2009, Standard for the Design of High-Performance Green Buildings, which specifies minimum requirements for the siting, design, construction, and planning for operation of high-performance green buildings.[8] The International Organization for Standardization has issued Standard 15392, Sustainability in Construction—General Principles.[9] It establishes internationally recognized principles for sustainability in building construction and establishes a common basis for communication of the information required among policy makers, regulators, manufacturers, building owners, and consumers. This standard may be especially useful to agencies with facilities outside the continental United States.

TECHNOLOGIES AND TOOLS

Many new technologies are available for use in new high-performance facilities and for retrofits of existing buildings. These technologies can be used to reduce greenhouse gas emissions; reduce the use of energy, water, fossil fuels, potable water, and toxic and hazardous materials; improve stormwater management; increase the use of renewable sources of energy; and take advantage of natural resources, including daylight, solar power, and geothermal.

The challenge of identifying the full range of high-performance systems and components that should be pursued for energy and environmental effectiveness is beyond the scope of this report. Key technologies for resource efficiency have been identified in reports such as *Real Prospects for Energy Efficiency in the United States* (NAS-NAE-NRC, 2010), *Greening Federal Facilities: An Energy, Environmental, and Economic Resource Guide for Federal Facility Managers and Designers* (DOE, 2001), and *Federal*

---

[7]Additional information is available at http://www.usgbc.org/DisplayPage.aspx?CMSPageID=1852#federal.

[8]Copies are available for purchase at http://www.techstreet.com/cgi-bin/detail?doc_no=ASHRAEI189_1_2009&product_id=1668986.

[9]Copies are available for purchase at www.iso.org.

Research and Development Agenda for Net-Zero Energy, High-Performance Green Buildings (NSTC, 2008), among others.

Examples of some of the technologies that should be of interest to federal agencies are the following:

- High-efficiency electrical lighting systems that incorporate state-of-the-art lamps, ballasts, and lighting fixtures; lighting fixtures that provide the desired lighting in the right places (e.g., task lighting); and controls that limit electrical lighting when daylighting is available and minimize heat loss.
- Fenestration (windows and doors) systems and designs that reduce heat gain in climates with high cooling requirements, while enabling daylight and views as well as passive solar heating when needed; low-E window glass to reduce heat loss.
- "Cool" roofs, including white roofs that offset the heating effect of carbon dioxide per unit area by reflecting incident sunlight and green (vegetated) roofs that retain rainwater and cool the building below.
- Solar technologies that can be incorporated into building roofs and facades for on-site power generation and water heating.
- HVAC controls that provide for the effective operation of the system during partial load conditions; split ventilation systems and thermal systems.
- Energy Star rated appliances and equipment that are more energy efficient than conventional appliances; WaterSense fixtures that are more water efficient than conventional fixtures; FEMP-designated electronic products that are more sustainable than conventional electronics; voice-over-Internet to replace standard telephones.
- Meters, sensors, and "dashboards" that provide real-time information on water and energy use.
- Porous pavers, cisterns, and low-flow irrigation systems to reduce the use of potable water and improve stormwater management; grey water systems that can recycle potable water, rainwater and stormwater for use in building equipment and for the landscape.

Where facilities occupy large, contiguous land areas, federal agencies have opportunities to install combined heat and power (co-generation) plants that can use a variety of renewable sources of energy (solar, wind, biomass, solid waste); arrays of solar panels and wind turbines; district energy systems and other technologies.

Most, if not all, of these technologies are most effective when used in combination with other technologies and when enabled by systems-based thinking, portfolio-based facilities management, integrated work processes, and other levers of change, as discussed below.

IDENTIFYING THE LEVERS OF CHANGE THAT ENABLE SPECIFIC TECHNOLOGIES AND SYSTEMS

Implementing the technological upgrades and advances that are needed in federal buildings to achieve energy and environmental goals may require specific levers of change. For instance, district energy systems and on-site power generation are possible with systems-based thinking and portfolio-based facilities management. District energy systems provide centrally managed supply and delivery of heating, cooling, and domestic hot water to concentrations of buildings in close proximity. Steam, chilled water, and hot water can be produced from both fossil fuels and renewable energy sources or a combination. District energy systems are closed-loop systems that are able to reuse heat that would be wasted in conventional open-loop systems. As previously pointed out, more than 65 percent of the energy produced in electricity generation at the power plant is lost in the form of heat.

District energy systems are widely used in Europe and Scandinavia and in some U.S. cities and on college campuses. Advanced systems use municipal solid waste, biofuels, and combined cycle solar and fuels for the combined generation of power and heat at the lowest energy cost.

Systems-based thinking combined with portfolio-based facilities management can also enable the use of grey water technologies to achieve more efficient use of potable water, the use of rainwater capture systems, and technologies for stormwater management such as porous pavers.

Integrated design processes are especially important in retrofitting buildings with new technologies because many technologies are interrelated and cross disciplinary boundaries. For example, energy use for lighting can be reduced by almost one-third by using T-8 lamps, electronic ballasts, occupancy controls, daylight dimming, and improved lighting design (NRC-NAE-NRC, 2010, p. 77). However, even greater savings are likely possible through integrated design of task ambient lighting with daylight through windows and skylights, task lights, and user controls of all three. Staff at the Lawrence Berkeley National Laboratory have calculated that average lighting energy use would be reduced from 1.69 to 0.45 watts per square foot of space through the introduction of an integrated office lighting system that combines lower levels of ambient overhead lighting with an efficient personal (task) lighting system (Brown et al., 2008).

Additional energy savings can be achieved from the increased use of daylight for ambient and task lighting. Indeed, daylight "harvesting" has been shown to reduce total lighting energy loads in buildings by 5 to 50 percent depending on the depth of the building (Rubinstein and Enscoe, 2010). The ultimate energy benefits are dependent on the performance of the integrated system of high-efficiency ambient and task electric lighting, daylighting through existing windows with light redirection, shade and glare control, as well as fully engaging the building occupants in determining appropriate light levels and maximum energy savings. The design of an integrated system is dependent not only on the architect, lighting engineer, and interior designer but also on the manufacturers of lighting fixtures, ballasts, lamps, and controls.

Integrated design processes are critical for meeting heating and cooling demands. A significant number of federal buildings are more than 40 years old (NRC, 1998), an age where perimeter heating systems and windows are failing and overall performance is declining, which increases the amount of energy used for heating, ventilation, and air conditioning. Perimeter heating and cooling demands are driven by climate and the performance of windows, so addressing these features together during building design and retrofit is important. Wall insulation and window technologies have advanced significantly in the past 10 years, and now have the potential to eliminate perimeter heating altogether, especially if internal heat gains can be used to offset perimeter losses through innovations such as advanced technologies for windows, including air flow windows. Replacing existing windows, heating, cooling, and air-conditioning systems that have reached the end of their service lives with more efficient systems, including automated control systems, can help to create higher-performance federal facilities. Energy reductions of as much as 55 percent in existing buildings could be realized through improved HVAC equipment and controls (NAS-NAE-NRC, 2010).

Portfolio-based facilities management can be a lever for improving the performance of roofs on federal buildings. Technologies for cool roofs, both white and green (vegetated) roofs (Figure 3.5), can contribute to reduced heating and cooling loads. In locations with summer cooling loads, both white and green roofs are equally effective at reducing the heat load on the building space below the roof. White roofs are almost three times as effective as green roofs when it comes to offsetting the heating effect of

FIGURE 3.5 White roof installed on the headquarters building of the Department of Energy; green roof installed on Chicago's city hall.

CO_2 (carbon dioxide) per unit area of roof, because of their greater solar reflectance. A notable advantage of green roofs, however, is that they retain up to 75 percent of incident rainwater, which subsequently evaporates, cooling the roof itself and the building below.

Although well-insulated white roofs can be ensured through performance-based procurement processes, systems-based thinking, combined with integrated work processes, can be used to develop cool, white roofs that also capture 100 percent of their rainwater to store and use on site, so they provide many of the advantages of green roofs. Rainwater capture for both types of roof typically involves installing a cistern at ground level and using the rainwater for irrigation or for equipment such as cooling towers. The commitment to on-site rainwater capture reduces the use of potable water for nonpotable uses, which can result in measurable energy savings and reductions in the use of hazardous chemicals for water treatment and transport.

These examples are intended to illustrate the importance of identifying the lever or set of levers most critical to enabling each technology or integrated design solution to rapidly advance the energy and environmental performance of federal buildings.

Examples of best practices that showcase the effective use of these technologies and others, as well as the other levers of change, are the focus of Chapter 4.

4

Best Practices, Tools, and Technologies for Transformational Change

To meet its charge, the committee was asked to identify current best practices and ways to optimize resources for achieving high-performance green building objectives during planning, design, construction, operations, and maintenance for new and existing facilities. As noted in Chapter 1, the committee defined best practices as "processes, procedures, or technologies that optimize available resources and could be effectively applied by the GSA and other federal agencies to meet similar objectives."

Chapter 4 highlights a range of practices, tools, and technologies that can be used to achieve a range of objectives for high-performance facilities. The best practices, tools, and technologies were primarily identified through presentations given to the committee at its meetings, the public workshop, and the workshop breakout sessions.

Summaries of some of the presentations are contained in Appendixes D through I. The summaries provide more context and detail about specific initiatives and should be read as an integral part of the committee's report. The presentations include the following:

- Transformative Action Through Systems-Based Thinking, by Bob Berkebile (Appendix D), discusses sustainable initiatives taking place in Greensburg, Kansas, North Charleston, South Carolina, and at several universities and also discusses the evolving nature of architecture and design.
- Sustainable Fort Carson: An Integrated Approach, by Christopher Juniper and Hal Alguire (Appendix E), focuses on the development of sustainability goals for Fort Carson, Colorado, and strategies and approaches being used to meet those goals.
- Beyond Incrementalism: The Case of Arlington, Virginia, by Peter Garforth (Appendix F), focuses on the collaborative, systems-based approach that is being used to develop a community-wide energy plan with the goal of more efficient use of energy and significant reduction of greenhouse gas emissions.
- Getting to Net-Zero Energy: NREL's Research Support Facility, by Jeffrey Baker (Appendix G), is a case study of a near-net-zero federal office building that was completed in June 2010 for the National Renewable Energy Laboratory (NREL).

- Sustainable Asset Management: The Case of Los Angeles Community College District (LACCD), by Thomas Hall (Appendix H), describes how an academic institution is creating nine campuses of high-performance facilities.
- The Economics of Sustainability: The Business Case That Makes Itself, by Greg Kats (Appendix I), presents evidence-based data from a new report, *Greening Our Built World: Costs, Benefits, and Strategies* (Kats, 2010).

BEST PRACTICES, TOOLS, AND TECHNOLOGIES RELATED TO SYSTEMS-BASED THINKING

The purpose of systems-based thinking is to find more efficient ways to use resources throughout their life cycle to deliver products and services. The following examples highlight collaborative processes for setting ambitious goals and learning how systems-based thinking is being used to achieve those goals. Tools to enable systems-based thinking, and technologies that are enabled by systems-based thinking are also discussed.

Examples of Collaborative Goal Setting

Greensburg, Kansas

Following the total destruction of their town by a tornado, the citizens of Greensburg, Kansas, used systems-based thinking as a basis for the town's reconstruction as a sustainable community (Figure 4.1). They systematically reconsidered the local economy, lifestyle, how people choose to use their time, and their future (see Appendix D). They developed a vision for what they wanted to achieve and then developed specific goals related to community, family, prosperity, environment, affordability, growth, renewal, water, health, energy, wind, and the built environment. Commitment from the whole community was sought at each stage of redevelopment planning and reconstruction (i.e., developing objectives and goals, preliminary design, detailed design, commitment of funds, construction, and postoccupancy).

By using systems-based thinking, the new Greensburg master plan optimizes the use of available resources. For example, rainwater and stormwater are captured in the landscape and the streetscape, purified, used, and then repurified for reuse (Figure 4.2). The sources of other resources, such as energy, were also identified and systematically evaluated to find ways to eliminate waste, optimize their use, and to achieve multiple objectives.

The Oberlin Project

The Oberlin Project is one of 18 Clinton Climate Initiative's climate positive projects. It grows out of Oberlin College's own campus sustainability initiatives and has become a collaborative venture involving Oberlin College, the municipal government, including the city schools, local townships, the municipal power company, private-sector organizations, local churches and nongovernmental organizations, and a major foundation. To help provide direction for the project, an advisory committee has been established that includes some of the nation's leading sustainability experts from architecture, urban design, renewable energy, and economic planning.

In contrast to Greensburg, Kansas, this project builds on an existing "town and gown" community. The goal is to transform Oberlin into a model of a post-fossil-fuel economy and sustainable development that can be widely emulated. Investments in building construction, renovation, and energy technology in a 13-block area of Oberlin's downtown are intended to fulfill multiple broad objectives. The objec-

FIGURE 4.1 Clockwise from left: Greensburg, Kansas, after tornado; new master plan; new, LEED-certified buildings. SOURCE: Top photo by Larry Schwarm. Other photos by BNIM Architects.

FIGURE 4.2 Conceptual design for new streetscape in Greensburg, Kansas. SOURCE: BNIM Architects.

tives include stimulating the expansion of existing businesses, creating new enterprises related to energy services, solar technologies, creating a vibrant arts community, and beginning a long-term conversation about the sciences around the many issues of sustainability (Figure 4.3).

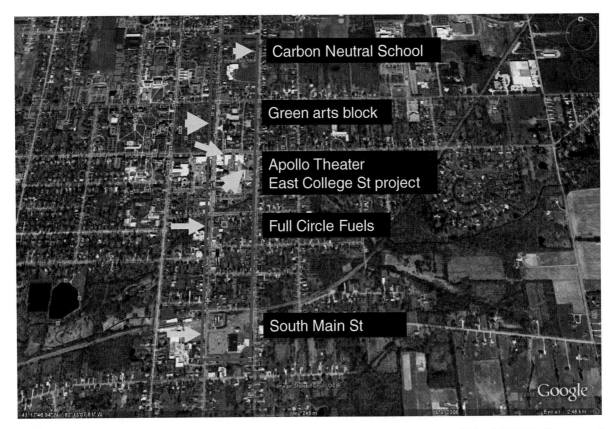

FIGURE 4.3 Thirteen-block area that is the focus of sustainable development in Oberlin, Ohio. SOURCE: Courtesy of David Orr.

Fort Carson, Colorado

Fort Carson, Colorado is one of the first three Army installations to pilot the concept of sustainability. The Fort Carson staff involved community leaders from the adjacent city of Colorado Springs to participate in the development of Fort Carson's original set of sustainability goals. The goals have since been updated (Box 4.1).

To implement these goals, the installation first used a hybrid management system that combined the aspirational goals with the existing environmental management system designed to ensure environmental legal compliance. The goals for sustainability have been integrated into the garrison's strategic plans, which are updated by the garrison commander to reflect multiple objectives related to soldiers, families, the workforce, and Fort Carson's training mission.

Personnel at Fort Carson are currently using systems-based thinking to achieve their transportation goals, including a 40 percent reduction in vehicle miles. They are working with local transit agencies and nonprofits to develop a regional transportation system that will serve the installation and the adjacent community. The staff has been able to develop innovative solutions by focusing on providing the service of mobility instead of focusing on existing infrastructure. The integrated mobility system now under consideration incorporates private-sector-provided car sharing, low-powered vehicle sharing (bikes, electric bikes), on-call transit services, enhanced telework strategies, and expansion of pedestrian and low-impact vehicle infrastructure (Figure 4.4).

BOX 4.1
Fort Carson Sustainability Goals 2002-2027, as of 2010

- 100 percent renewable energy, maximum produced on the installation
- 75 percent reduction of potable water purchased 2002-2027
- Sustainable transportation achieved, characterized by 40 percent vehicle miles reduction from 2002 and development of sustainable transportation options
- Sustainable development (facilities planning)
- Zero waste (solid waste, hazardous air emissions, wastewater)
- 100 percent sustainable procurement
- Sustaining training lands—meaning ongoing capability of the biological health of training lands in support of the installation's primary mission to train soldiers

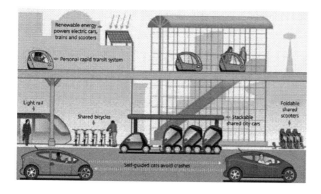

FIGURE 4.4 Fort Carson Concept for multi-modal solar powered sustainable mobility. SOURCE: Courtesy of PRT Consulting (Appendix E).

The staff continues to use collaborative processes to achieve a variety of goals. For example, they are currently developing a regional sustainable energy plan, the Pristine Energy Project, in partnership with the Pikes Peak Sierra Club. Technical support is being provided as needed on a volunteer basis by NREL, the Colorado Governor's Energy Office, the Southwest Energy Efficiency Project, and the Colorado Renewable Energy Society. One aspect of the planning process is to identify the barriers to providing renewable sources of energy to the customers who want it. It is anticipated that a plan will be available in 2011 that will identify for public policy makers a path for helping buyers who want renewable energy to be able to buy it from providers at a reasonable cost (Appendix E).

Arlington County, Virginia

This close-in suburb to Washington, D.C., is developing a community energy plan around the goals of competitiveness, security, and the environment. This is a collaborative effort of the Arlington County government, representatives from the Chamber of Commerce, local gas and electric utilities, property developers, civic associations, and major governmental landholders, including Ronald Reagan Washington National Airport, the Pentagon, and Fort Myer. A community task force was established to provide input and oversight for the effort. In addition, a technical working group was established that included experts from North America and Europe to provide wide-ranging expertise and differing perspectives.

In developing the plan, the community established goals. The environmental goal is for the reduction of greenhouse gas emissions at a breakthrough level. Systems-based thinking is being used to identify where and how electricity is produced, how it is transferred to the county, and how much is used by businesses, government, residents, commuters, and others. This type of analysis will serve as the basis for determining where there are opportunities to reduce energy use by implementing district energy systems, public transit, and other solutions. This effort takes advantage of the results and lessons learned from community energy plans implemented in Copenhagen, Denmark, and Guelph, Canada, which have documented their results (Appendix F).

Many additional systems-based initiatives are under way that deserve study. They include programs at the Army's Fort Lewis, Fort Hood, and Fort McCord installations, the Department of the Navy's Energy Program for Security and Independence,[1] and Cornell University's Climate Action Plan.[2]

Tools for Enabling Systems-Based Thinking

Available and emerging Internet-based tools can be used by federal agencies to support systems-based thinking and life-cycle assessment and to evaluate the environmental impacts of building materials and components. Such software programs help to evaluate upstream decisions on purchasing and the supply chain for materials, energy, water, and other resources. The following were identified during the course of this activity:

- *Building for Environmental and Economic Sustainability (BEES).* Developed by the National Institute of Standards and Technology Engineering Laboratory, the BEES software allows for the selection of 230 cost-effective, environmentally preferable building products. All stages in the life of a product are analyzed: raw material acquisition, manufacture, transportation, installation, use, and recycling and waste management. Economic performance is measured using life-cycle costing, including the costs of initial investment, replacement, operation, maintenance and repair, and disposal. Environmental and economic performance are combined into an overall performance measure using the American Society for Testing and Materials standard for Multi-Attribute Decision Analysis. Additional information is available at http://www.nist.gov/el/economics/ BEESSoftware.cfm.
- *The Athena Institute.* A nonprofit organization, this institute has developed life-cycle tools for evaluating the sustainability of building materials, building assemblies. and whole buildings. Additional information is available at http://www.athenasmi.org/index.html.
- *One Planet Communities* incorporates a systems-based approach for planning at the community, campus, or portfolio level. Originally developed in the United Kingdom, the system establishes 10 areas of study and corresponding goals. Software is available to help with the assessment of possible approaches during the design alternatives phase. Additional information is available at http://www.oneplanetcommunities.org/about-2/principles/.
- *EARTHSTER.* This software program is scheduled for release in the fall of 2011. The development of EARTHSTER began during the design of a new school of nursing in Houston, Texas, when it was determined that a more efficient process was needed to analyze best material choices. Software was created to allow the designers to use a large body of data collected by the Environmental Protection Agency and the Department of Commerce for counties. The data were used to evaluate the upstream environmental impact of design decisions and then to improve the selections and the

---

[1]Available at http://greenfleet.dodlive.mil/files/2010/04/Naval_Energy_Strategic_Roadmap_100710.pdf.
[2]Available at http://www.sustainablecampus.cornell.edu/climate/.

performance of the building and to evaluate the economic consequences of decisions (Appendix D). The current version of EARTHSTER is being piloted by Walmart to communicate with and improve the performance of its suppliers. When the system is released, it will be open and free to all users who are willing to contribute information about their supply chains. The ultimate goal of the software designers is to provide actionable analysis of the environmental and social impacts of product life cycles and supply chains to accelerate the transition to sustainable products. Additional information is available at www.earthster.org.

BEST PRACTICES, TOOLS, AND TECHNOLOGIES RELATED TO PORTFOLIO-BASED FACILITIES MANAGEMENT

Well-designed facilities portfolio management programs start with a clear framing of facilities-management goals linked to overarching organizational goals and mission and a careful blueprint for the capabilities and requirements of existing facilities. Examples of best practices for goal setting and for planning and programming are identified below. Because the performance of existing buildings will be critical if agencies are to meet their goals for high-performance facilities, efficient operations and replacements and retrofits of existing building systems will be especially important.

Goal Setting

Fort Carson and the Office of Overseas Buildings Operations provide examples of best practices for setting goals that tie both facilities management and sustainable strategies to organizational mission:

- *Fort Carson, Colorado*. The goal of the facilities management team is to provide mission support and services, including quality of life programs for Fort Carson soldiers, families, and community. To enable the organizational mission, they have adopted a sustainable approach for facilities that provides superior work and living environments for soldiers and their families and avoids facilities-related costs by saving energy, water, and other natural resources.
- *State Department's Office of Overseas Buildings Operations (OBO)*. The goal of the OBO is to create platforms for eco-diplomacy by greening U.S. embassies and consulates worldwide and to minimize the impacts of overseas facilities by designing, constructing, operating, and demolishing buildings in an energy-efficient and environmentally sensitive manner.

Planning and Programming

Examples of effective planning and programming practices using a portfolio-based approach were the following:

- *Virtual embassies*. Where appropriate, the OBO is minimizing the environmental impact of its overseas facilities by providing services to citizens and others online instead of building new embassies. By doing so, they have eliminated the need for capital investment, avoided building-related life-cycle costs, mitigated the environmental impacts of buildings, reduced travel to and from an embassy, and improved the physical security of State Department employees by not placing them in harm's way. OBO staff acknowledge that an online presence is not appropriate for all countries or situations. However, where appropriate, the cost savings and other benefits that accrue to the State Department and others from virtual embassies are likely to be substantial. One example is the virtual consulate for Bangalore, India (http://bangalore.usvpp.gov).

- *Historic buildings.* Federal agencies manage historic buildings as part of their portfolios. Many historic buildings were high-performance green buildings when they were originally constructed: They used passive techniques for heating and cooling coupled with natural daylighting and ventilation strategies. However, their performance may have been compromised over time through the accretion of mechanical systems and the elimination of original components. By carefully retrofitting and replacing existing systems, some historic buildings can become high-performance buildings again. One example of such an effort is the 92-year old Wayne Aspinall federal building in Grand Junction, Colorado. The GSA is contracting to rebuild the structure from within by replacing existing mechanical and electrical systems. The goal is to create a zero-net-energy building, which could make it the first zero-net-energy building on the National Register of Historic Places (Daily Sentinel, December 2, 2010) (Figure 4.5).

Operations, Maintenance, Repair, and Replacement

Federal agencies have significant opportunities to upgrade the performance of existing building systems through effective operations, through routine maintenance, repair, replacement programs, and through retrofit projects.

Building commissioning is a well-recognized best practice for effective operation of building systems in existing conventionally designed buildings as well as newer buildings designed to be more sustainable. It is intended to ensure that building systems are installed that perform at the level to which they were designed. If building systems are operated and maintained to continue to perform at that level, the result will be lower energy consumption and lower life-cycle costs than if the systems are not operated and maintained appropriately.

Total building commissioning is an overarching process that can be effectively used for new buildings beginning in the design phase, including the setting of goals related to design for operability and commissioning at the start of the project. The Guiding Principles for Federal Leadership in High Performance and Sustainable Buildings recommend the use of total building commissioning practices in order to verify the performance of building components and systems and help ensure that design requirements are met.

For existing buildings, the recommissioning of heating, ventilation, air conditioning (HVAC) systems can be undertaken every 3 to 5 years to ensure that mechanical systems are able to deliver thermal comfort and air quality in an energy-effective manner. Studies published by the Lawrence Berkeley

FIGURE 4.5 Wayne Aspinall Federal Building, Grand Junction, Colorado. SOURCE: GSA.

National Laboratory identified 15 percent energy savings through HVAC commissioning, with less than a 9-month payback, across a large portfolio of existing buildings (Mills, 2009).[3] The city of New York is incorporating building commissioning as an important strategy for creating significant energy and environmental impacts into its planning documents.[4]

Because different building components wear out at different rates, there is an opportunity to replace worn-out components with more efficient ones on a routine basis. As water and light fixtures, appliances, or computers wear out or become obsolete, they can be replaced by WaterSense fixtures,[5] compact florescent lamps, Energy Star equipment and appliances, and the like.

Additional significant improvements in the performance of existing buildings can be achieved through the integration of technologies when roofs, windows, heating, and lighting systems are replaced or during major retrofits. When retrofits are undertaken, retro-commissioning or recommissioning processes can be used to ensure that the upgraded mechanical systems are installed to operate as they were designed or to ensure the airtightness of building enclosures and insulation.

In any given year, federal agencies replace the roofs on a significant number of buildings. Replacing conventional black bitumen roofs, which typically wear out after 20 years, with cool white roofs or vegetated green roofs, which have 20-30 year service lives, can be cost effective. Both white and green roofs have additional advantages over conventional roofs and in relation to each other, as described in Chapter 3.

Tools to Enable Portfolio-Based Facilities Management

Many federal agencies already use computerized maintenance management systems and other tools to support the management of their operations and maintenance programs.

The EPA's *Portfolio Manager* is an interactive energy management tool that allows an organization to track and assess energy and water consumption across its entire portfolio of buildings in a secure online environment. It can help an organization set its investment priorities, identify underperforming buildings, and verify efficiency improvements. Additional information is available at http://www.energystar.gov/index.cfm?c=evaluate_performance.bus_portfoliomanager.

Technologies Enabled by Portfolio-Based Facilities Management

Several organizations described initiatives using district energy systems and combined heat and power plants. District energy systems are especially well suited to military installations, campus-type settings, and areas of higher density development, such as portions of cities. As part of its community energy plan, Arlington County, Virginia, is studying how it can develop district energy systems to service existing concentrations of high density development (Appendix F).

The Los Angeles Community College District (LACCD) is creating district energy systems and testing a number of renewable energy sources and generation and storage technologies across its nine campuses (Appendix H). One plant incorporates solar thermal technology to take care of the heating and cooling load of a campus by using an absorption chiller for cooling and stored hot water for heating. A second campus is using a combination of solar power for energy generation and thermal energy storage in the form of both ice storage and hot water storage (Figure 4.6).

---

[3]Report available at http://eetd.lbl.gov/ea/emills/presentations/mills_cx_ucsc.pdf.
[4]Additional information is available at www.nyc.gov/planyc2030.
[5]Additional information is available at http://www.epa.gov/WaterSense/.

FIGURE 4.6 Valley College central plant components: Solar array (*top left*), hot water storage (*top right*), new infrastructure for delivering power; and vacuum tube heat-pipe collectors (*bottom*). SOURCE: Courtesy of the LACCD.

BEST PRACTICES, TOOLS, AND TECHNOLOGIES RELATED TO INTEGRATED WORK PROCESSES

Integrated work processes can be used to overcome fragmented decision making and are essential for achieving the multiple objectives associated with high-performance facilities. Such processes incorporate extensive up-front planning involving all significant stakeholders in order to optimize the choice of materials, energy systems, and other building components.

The research support facility at the NREL (Appendix G) exemplifies the effective use of integrated work processes, from goal setting through design and acquisition. The resulting building integrates the use of the locally available natural resources with building technologies to create an office building that will be highly energy efficient.

NREL Research Support Facility

The NREL staff developed three sets of performance-based goals for the proposed building and established a fixed budget for the building, its furnishings, and equipment. The goals included a set of energy requirements, space for 800 people, and achievement of a LEED Platinum rating.

An integrated acquisition process was used. The performance-based goals allowed the contractors bidding on the project to use their creativity in providing alternative designs that would meet the goals. Components of the acquisition strategy included a performance-based request for proposals, a national conceptual design competition, design-build project delivery, and a firm-fixed price contract with incentives.

Throughout the project, the owner and the designers analyzed all aspects of the new building's energy use, including equipment, the data center, and lighting. They considered what natural resources were available to provide lighting, heating, and cooling in order to reduce the energy used by electrical and mechanical systems. They also looked at total building energy performance over its life cycle, including how the building would be operated.

The energy budget and design for the project

- Included total energy use that would take place in the building from appliances, equipment, computers, and the like;
- Required replacing all computers and electronic systems with more efficient ones;
- Provided 100 percent day lighting of all work spaces;
- Used the structural concrete foundation system as a large thermal battery for free heating and cooling from the outside; and
- Included full monitoring of energy use.

The opportunity to integrate the foundation into the heating and cooling system could only be realized through effective up-front planning: Retrofit after the fact would not have been possible. NREL staff acknowledged that the process required significant owner involvement up front. However, the resulting building met or exceeded the established goals and was also completed within the original budget and at the same cost per square foot as a conventional building in that region.

The resulting building is shaped to take advantage of its climate and integrates a range of progressive technologies. The southern facade incorporates careful solar shading and transpired solar collectors—dark-colored perforated metal sheeting that preheats incoming air and stores it in the building's crawl space—along with waste heat recovered from the computer data center. The building itself—its walls, floors, and foundations—functions as a "large thermal battery," storing and then releasing free heat or extracting it in cooling mode, boosted by radiant floor heating and evaporative cooling if required. Photovoltaic arrays have been installed across the building's roof and the adjacent visitors' parking lot through a power purchase agreement.

Workspaces in this building are almost 100 percent daylit—a function of the building's narrow (60 foot) floor plate and the incorporation of "light-shelves" above the windows that reflect daylight onto the ceiling. By capturing daylight and eliminating the waste of heat by electric lighting, the designers were able to reduce the size and cost of mechanical and electrical systems and invest those savings in a higher-performance façade (Figure 4.7).

To reduce electricity use, desktop computers were replaced with laptops, standard telephones were replaced with Voice-over-Internet, and a system was installed that turns the power off if a computer is not in use.

Tools to Enable Integrated Work Processes

Having quality information available at the beginning of a project can support effective decision making about the design of new buildings or retrofits and can improve the outcomes significantly. A range of modeling, virtual design, and other technologies are available to support integrative design processes. Such applications are particularly powerful when combined in interoperable models (often referred to as building information models, BIMs) that allow for the sharing of electronic data among

FIGURE 4.7 *From top left:* NREL research support facility; labyrinth thermal storage; daylit interior. SOURCE: U.S Department of Energy, National Renewable Energy Laboratory, and Pat Corkery.

a project's owners, clients, contractors, and suppliers, and across an organization's design, engineering, operations, project management, financial, and legal units (NRC, 2009).

Tools to enable integrated design processes include digital terrain models and civil features; architectural models for façade, roof, and interior; structural models for analysis, design, and detailing; energy models for equipment, plug loads and lighting; and integrated BIMs for analysis and coordination of all types of building systems. Integrated modeling tools are also beginning to allow partially automated fabrication of materials and components for high-performance buildings, such as earthwork and paving (using GIS), structural and reinforcing steel, building facades with complex geometry, process and services piping, and HVAC ductwork.

During the planning for the NREL research support facility, energy models were used extensively: Every design decision was checked against the energy model, a practice that was critical to the design outcomes, particularly when it was necessary to make trade-offs (Appendix G). Similarly, models were used to test drive the design of the University of Georgia's new Odum School of Ecology against two existing buildings to determine its performance in energy and water use (Appendix D).

BIM technologies were effectively used in the design of the Internal Revenue Service building in Kansas City, Missouri, to analyze many scenarios quickly so as to advance the best concepts—those that would satisfy a range of stakeholders and create an environment in which the occupants could feel good about their environment and experience better health and increased productivity (Appendix D).

The design of the Point Pavilion band shell in North Charleston, South Carolina, exemplifies another use of BIM (Figure 4.8). In this case, the deadline for completion of the project did not allow for conventional design and construction practices. The architects e-mailed the design (using BIM) to the contractor. The design documents were created and approved electronically and then entered directly into the contractor's manufacturing system (computer-controlled fabrication) (Appendix D).

FIGURE 4.8 Point Pavilion designed and delivered using BIM. SOURCE: BNIM Architects.

The committee recognizes that a number of barriers remain in making interoperability and building information modeling a fully operable, deployable technology. First, the use of BIM applications varies significantly among architects, engineers, general contractors, and subcontractors (Jones, 2009). The applications and technologies are only rarely integrated across all phases of a project, so their benefits are not fully optimized. In addition, barriers remain in developing fully operable systems, including legal issues, data storage capacity, and the ability to search thousands of data items quickly to support real-time decision making. Because there is insufficient interoperability within the capital facilities sector of the construction industry, causing $15.8 billion in inefficiencies and lost opportunities every year (NIST, 2004), development should continue as rapidly as possible (NRC, 2009).

BEST PRACTICES, TOOLS, AND TECHNOLOGIES RELATED TO PROCUREMENT AND FINANCE

Because funding levels will likely remain unchanged or be reduced in future federal budgets, federal agencies will need to find ways to leverage their available resources if the goals and objectives for high-performance facilities are to be achieved. The committee identified a range of best practices for procurement and third-party financing of projects that could be used by federal agencies. Among these were more efficient procurement practices, collaborative partnerships, and the strategic use of volume purchasing power.

Efficient Procurement Practices

More efficient procurement practices have the potential to significantly reduce transaction costs and save staff time. For example, to save time and money in procuring products for its campuses, the LACCD has developed master agreements. All nine colleges work from the same agreement to procure furniture and there is a specified standard for recycled material, carpeting, and other products. Such agreements cut transaction costs by improving efficiency: It is no longer necessary for each campus to develop and confirm its own set of specifications. An additional outcome is that such agreements allow the LACCD to take advantage of its purchasing power to secure discounts from providers who know that the LACCD will be buying in large volumes (Appendix H).

Guidelines can also take the form of checklists with questions that can be applied at every decision point to help embed sustainability at all levels of decision making. An example of such a checklist to promote sustainable spending decisions is shown in Box 4.2.

To be able to use solar power on a widespread basis, the State Department procures solar panels centrally and then distributes them to its embassies. The provider of solar panels is Unicor, which is operated by Federal Prison Industries as a self-sustaining, self-funded corporation. The Federal Prison

---

BOX 4.2
Illustrative Guidelines: Operating Discipline for Smart Spending

- If you do not need it, do not buy it.
  —Will buying this product or service contribute to my ability to meet/improve customer/client offering or maintain a safe environment?
  —Can I use less?
  —Can I source it within the organization?

- Understand if the cost and value of what you think you require is more than the minimum or standard requirement.
  —Will the customer/client see the value in the higher specification and pay for it?
  —Will the minimum still ensure a safe environment?

- Plan the use or consumption of goods and services.
  —Minimize inventory by buying just in time rather than just in case.
  —Avoid reactive or emergency purchases.
  —Take time to look for the best value.
  —Include procurement in production and maintenance planning and scheduling, in shutdown preplanning, and in capital projects.

- Optimize the buy.
  —Include sourcing professionals in the process.
  —Pay only for value-in-use.
  —Maximize use of standardization (generic brands).
  —Leverage volume for price by using converged suppliers.
  —Have supplier maintain inventory.
  —Improve payment terms as well as price.

- Include the supplier in your work simplification and improvement processes. Get alignment for win-win. Completely maximize the total value the supplier has to offer.

SOURCE: Adapted from DuPont Company.

Industries was established in 1934 by Executive Order to create voluntary real-world work programs to train federal inmates.[6]

Agencies regularly contract with private-sector firms to provide design and construction services for new buildings and major retrofits. The use of performance-based contracts allows agencies to set high-level goals and then challenge private-sector firms to use their creativity and knowledge of the sustainable practices to meet those goals. The integrated acquisition process used by NREL for its research support facility was previously described.

At Fort Carson, all new major construction projects are being LEED certified through the U.S. Green Building Council. Contracts for new building construction require the design-builder to submit an implementation plan that includes the following:

- An air quality plan,
- A waste management plan,
- A commissioning plan,
- A LEED schedule,
- A personnel role list,
- A 500-mile radius map to show where the materials will come from, and
- A narrative on how every criterion for meeting LEED requirements will be achieved.

For one of its new buildings, a rating of LEED Silver was the original goal. The contractor decided to work toward additional points so that the building could be certified as LEED Gold. The contractor can now use this accomplishment—designing and building the first LEED Gold building in the U.S. Army—when competing for future work.

Fort Carson's source selection boards look for contractors with past experience in LEED projects. Fort Carson's has four LEED-accredited professionals on staff as of FY 2011.

Several potential best practices were identified in the workshop breakout sessions to address various contracting issues. One addressed the issue of time lag and project designs going stale, such that the project will not be state of the art when the ribbon is cut. In these circumstances, agencies could work with contractors through charrettes or other practices to update the designs to state-of-the-art standards before construction. A second suggestion was to make reduced operations and maintenance costs an evaluation selection criterion when soliciting proposals and selecting contractors for new and retrofit projects. In this way, contracts could help incorporate a life-cycle perspective. Additional suggestions were to create government-wide centers of excellence for business-based core contracting processes (these could be virtual) and to develop model clauses and best-practice procedures for performance-based contracts that could be used by all federal agencies.

Finance

With already constrained budgets and the likelihood of future cuts, federal agencies will need to leverage their available funding through public-private partnerships. Energy Savings Performance Contracts (ESPCs) are one type of public-private partnership that many federal agencies are already using. Approximately $2.3 billion has been invested in federal facilities through ESPCs to achieve a savings of 18 trillion Btu approximately equivalent to the energy used by a city of 500,000 people. The ESPC

---

[6]Additional information available at www.unicor.gov.

projects contain guarantees that will result in $6 billion in avoided energy costs over the life of the contracts (Kidd, 2010).[7]

Under such agreements, an energy service company (ESCO), such as a utility, typically conducts a comprehensive energy audit for federal facilities and identifies improvements to save energy. In consultation with the agency that owns the facilities, the ESCO designs and constructs a project that meets the owner's needs and arranges the necessary financing. It guarantees that the improvements will generate energy cost savings sufficient to pay for the project over the term of the contract. After the contract ends, all additional cost savings accrue to the owner organization.[8]

The use of ESPCs does have its limits. At the workshop some suggested that agencies could realize greater savings if funding were directly appropriated for this purpose so that agencies could implement the energy retrofits directly and not involve a third party. One presenter questioned whether the narrow focus of ESPCs was actually hindering the achievement of a broader range of goals. He also questioned whether the ESCOs were only choosing those projects that were the easiest to implement as opposed to those that might be more difficult and yield lower profits (Appendix I). The State Department's Office of Overseas Buildings Operations has moved to centralized procurement of solar panels because they found it difficult to use ESPCs across countries and locations due to the great variation in local cultures and capabilities.

Power purchase agreements (PPAs) are another public-private finance arrangement being used. These agreements allow an owner organization such as a federal agency to finance on-site renewable energy projects with no up-front capital costs incurred. With a PPA, a developer installs a renewable energy system on the owner organization's property under an agreement that the organization will purchase the power generated by the system. The organization pays for the system through these power payments over the life of the contract. After installation, the developer owns, operates, and maintains the system for the life of the contract.[9]

At Fort Carson, a power purchase agreement with a private-sector firm that would develop a wood biomass co-generation facility on Fort Carson property is under consideration. The Front Range Energy Consortium, made up of five Air Force and Army military installations, is investigating the potential for developing a 50-megawatt concentrated solar installation at an Army chemical depot site in Pueblo, Colorado (Appendix E). NREL used a power purchase agreement for its photovoltaic arrays.

To help develop a regional mobility system, staff at Fort Carson are working with local transit agencies and nonprofits to see how they might leverage federal employee mass transportation benefits to support new local public transportation options (Appendix E).

Revolving funds were identified at the workshop as a potential financing mechanism for green projects. A specific example is Harvard University's Green Campus Loan Fund that provides up-front capital for projects that reduce environmental impacts and have a payback period of 5 years or less (http://green.harvard.edu/loan-fund).

Driving Markets

Executive Order 13514 directs federal agencies to drive the market for sustainable products. An example of how this could be done was provided by the LACCD, which was able to change the market for carpeting. The LACCD hired an expert in carpets to write the specifications for a sustainable carpet that could be recycled at the end of its service life. Once the specifications were written, the LACCD

---

[7]Additional information on ESPCs is available at http://www1.eere.energy.gov/femp/financing/espcs.html.

[8]Source: Federal Energy Management Program Web site at http://www1.eere.energy.gov/femp/financing/espcs.html.

[9]Source: Federal Energy Management Program Web site at http://www1.eere.energy.gov/femp/financing/power_purchase_agreements.html.

worked with manufacturers to develop a procurement process that would guarantee a market for the product if the manufacturers actually produced it. The carpet mills changed the way they were manufacturing carpet. The LACCD now has a more sustainable carpet that costs 50 percent less over its life cycle, saving both capital and operational funding. And the carpet mills have a new product that is being sold on the open market for a profit (Appendix H). In the state of California, similar types of collaborative arrangements have been used to spur the manufacture of compact fluorescent lamps.

Agencies can also drive markets through their leasing standards, and Executive Order 13514 includes provisions for doing this. The recent report of the President's Council of Advisors on Science and Technology recommended that federal agencies be given authorization to enter into ESPCs for leased facilities (PCAST, 2010). Federal agencies could also help drive the market for the continued development of BIM and other interoperable applications by requiring contractors to use these technologies for projects related to new construction or major retrofits.

During the study, it was suggested that the benefits of Energy Star labeling could be increased by setting up tiers of Energy Star labels—double and triple stars—to move beyond 20 percent better than conventional standards and to establish top 10 percent and top 5 percent levels. The federal government could then drive the market by procuring only triple-star appliances and fixtures. Energy Star labeling could also be extended to other appliances and fixtures that account for major energy costs for the federal sector, including flat screen monitors, vending machines, chargers, and office kitchen equipment.

BEST PRACTICES, TOOLS, AND TECHNOLOGIES RELATED TO COMMUNICATION AND FEEDBACK FOR BEHAVIORAL CHANGE

Effective communication and feedback that help to spur cultural and behavioral change can take many forms. A range of best practices, tools, and technologies that could support such change were identified at the workshop and in presentations to the committee, as described below.

Examples of Communication and Feedback for Behavioral Change

Fort Carson, Colorado

After involving local community leaders in the development of sustainable goals for Fort Carson, staff have continued to communicate regularly with community leaders. Annual conferences to report progress on meeting the sustainability goals have been held with the community. The garrison commander also hosts monthly sustainability breakfasts with Fort Carson staff and community leaders to discuss long-term goals and issues, to generate potential solutions, and to provide for long-term engagement in the achievement of the sustainability goals. These types of activities help to generate excitement and energy among all of the stakeholders over the long term (Appendix E).

Fort Carson also provides awareness training for all soldiers and employees and competence training for managers that includes integration of sustainability performance with the installation's strategic plans.

Branding of Sustainable Initiatives, Buildings, and Products

The purpose of branding is to differentiate a project, initiative, or product from others in the market and influence the consumer such that he or she will want to buy or buy into that product. Rating systems such as LEED, BREEAM, Green Globes, and Energy Star, are examples of branding that have been used to change the behavior of consumers.

Two additional examples of branding identified at the workshop were the following:

FIGURE 4.9 Fort Carson Logo. SOURCE: Fort Carson, Colorado.

- To better reach the 150,000 people who regularly use its installation and its 30,000 on-site personnel, the Fort Carson staff developed a logo in 2010 (Figure 4.9). The logo is used to provide consistent and modern messaging techniques and to encourage people to want to be part of the brand (Appendix E).
- One of the strategies under consideration in the draft Community Energy Plan of Arlington County, Virginia, is the use of energy performance labels for buildings. An energy performance label would be available whenever a building is sold or rented. It would typically be prominently displayed in buildings regularly used by the public. Actual energy performance would be independently certified. The specific labeling approach has not yet been defined but will probably be similar to the emerging ASHRAE Energy Quotient approach,[10] which in turn is an adaptation of the European Union performance labels[11] (Appendix F).

Evidence-Based Data

To help change behaviors, facility managers need to be able to present evidence-based data (factual information based on measured results) to decision makers, operators, and occupants demonstrating that increasing funding for design and construction of a project by 1 or 2 percent can result in substantial long-term savings as well as cost avoidances. In addition to the previously cited report (Kats, 2010), at least one other detailed analysis of the economic value of the certification of green office buildings has been published by the University of California-Berkeley (Eicholtz et al., 2009).[12]

The information generated through energy audits and other reporting requirements can be used to communicate with decision makers, building operators, and occupants about successes and inefficiencies and about ways to further improve efforts aimed at energy reduction or other goals.

NREL Research Support Facility

The NREL staff recognized that occupant behavior would be essential for the ultimate performance of the new research support facility. They established a set of guidelines for occupants that prohibit the use of individual coffee pots or space heaters.

---

[10]American Society of Heating, Refrigerating, and Air-Conditioning Engineers, "ASHRAE Introduces Prototype of Building Energy Label at Annual Conference," June 22, 2009, at http://www.ashrae.org/pressroom/ detail/17194.

[11]EurActiv, "Energy Performance of Buildings Directive," September 29, 2010, at http://www.euractiv.com/en/energy-efficiency/energy-performance-buildings-directive-linksdossier-188521.

[12]The paper "Doing Well by Doing Good? Green Office Buildings" is available at http://escholarship.org/uc/item/507394s4.

FIGURE 4.10 Space layout at NREL research support facility. SOURCE: Department of Energy, National Renewable Energy Laboratory.

The NREL staff also recognized that cultural change would be necessary but could be difficult to achieve. One of their biggest cultural challenges was furniture: Staff would be moving from existing offices with hard walls and private offices to a new building with lower walls and cubicles that optimize daylight. To achieve acceptance of this change on the part of staff, a test office was set up. Staff used the test office for a year and a half. Through that process, NREL staff worked with the furniture manufacturers to improve the layout before the staff moved into the new building (Figure 4.10).

Tools and Technologies to Enable Behavioral Change

A variety of tools and technologies can be used to enable performance measurement, communication, and behavioral change. One of these, the use of metering in buildings to better track performance, is already under way: Federal agencies are required to meter their buildings by October 1, 2012. The use of submetering, a more refined tracking of energy and water use that can be applied to multiple tenant buildings, is also being studied. Displaying real-time meter interval data at locations that are easily accessible to building occupants and the public is one way to support behavioral change, as illustrated in the following examples.

Oberlin College initiated development of a campus resource monitoring system (first-of-its-kind technology) that provides students with real-time feedback on their electricity and water use in dormitories to engage, educate, and empower them to conserve resources. The result is an ongoing competition among students, pressure on the university to improve the energy and water performance of some of their buildings, and a measurable reduction in energy and water demands over the past 10 years.

At the workshop, it was reported that some agencies have set up competitions among the occupants of different buildings or different divisions to see which ones can achieve the greatest reductions in energy use, so-called "biggest losers" programs. Reminding employees to turn off computers and other equipment and use energy-saving features could also result in reductions in energy use, which might be boosted by setting up competitions.

At NREL, a variety of technologies were used to reduce and monitor energy performance in ways that were clear to occupants and helped to change their behavior (Figure 4.11).

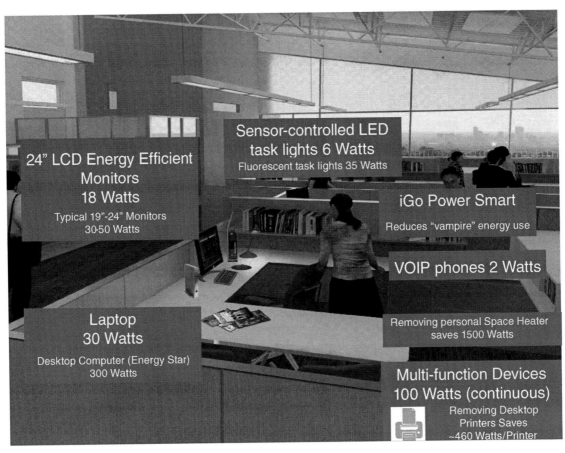

FIGURE 4.11 Technologies used to reduce energy use and change occupant behavior. SOURCE: Jeffrey Baker, Department of Energy, National Renewable Energy Laboratory.

5

Strategies and Approaches for Achieving a Range of Objectives Associated with Federal High-Performance Facilities

In Chapter 5 the committee recommends 12 strategies and approaches that the GSA and all federal agencies can use to achieve a range of objectives associated with federal high-performance buildings and facilities. The strategies and approaches are based on the levers of change, the best practices, tools, and technologies identified at the public workshop and other meetings, and on the committee members' own expertise. They are intended to optimize the use of natural, financial, and human resources and to minimize environmental impacts. The recommended strategies can be applied by federal agencies to their portfolios of facilities as well as to individual building projects.

OBJECTIVES ASSOCIATED WITH FEDERAL HIGH-PERFORMANCE GREEN BUILDINGS

The Energy Independence and Security Act (EISA) of 2007 defined the attributes of a federal high-performance green building. Taken as a whole EISA, Executive Order 13423, *Strengthening Federal Environmental, Energy, and Transportation Management,* and Executive Order 13514, *Federal Leadership in Environmental, Energy, and Economic Performance*, establish more than 20 objectives related to federal high-performance facilities, including the following:

- Reducing the use of energy, potable water, fossil fuels, and materials;
- Reducing greenhouse gas emissions;
- Improving indoor environmental quality;
- Increasing the use of recycling and environmentally preferable products;
- Minimizing waste and pollutants through source reduction;
- Pursuing cost-effective innovative strategies to minimize consumption of energy, water, and materials;
- Leveraging agency acquisitions to foster markets for sustainable technologies, materials, products, and services;
- Locating new buildings in sustainable locations;

- Participating in regional transportation planning;
- Strengthening the vitality and livability of the communities in which federal facilities are located;
- Eliminating fossil fuel energy use in new buildings and major renovations by 2030; and
- Beginning in 2020 and thereafter, designing all new federal buildings to achieve zero net energy by 2030.

Each mandate specifically calls for the use of a life-cycle perspective or life-cycle costing, establishes interim and longer-term targets for the objectives, and establishes baselines and performance measures for evaluating progress in achieving them.

STRATEGIES AND APPROACHES FOR ACHIEVING A RANGE OF OBJECTIVES RELATED TO FEDERAL HIGH-PERFORMANCE FACILITIES

The strategies and approaches are summarized in Box 5.1. More detailed explanations for each one follow.

1. Use systems-based thinking and life-cycle assessment to identify new ways to provide services and to eliminate waste.

BOX 5.1
Summary of Strategies and Approaches for Achieving a Range of Objectives Related to Federal High-Performance Facilities

1. Use systems-based thinking and life-cycle assessment to identify new ways to provide services and to eliminate waste.
2. Focus on community- and regional-based approaches to fill gaps, leverage resources, and optimize results.
3. Align existing federal facilities to current missions and consolidate the total facilities footprint to lower costs, reduce carbon emissions, reduce water and energy use, and optimize available resources.
4. Operate facilities efficiently to optimize their performance.
5. Aggressively implement proven sustainable technologies as a matter of course.
6. Use integrated, collaborative processes and practices to overcome conventional segmented processes that fail to optimize resources.
7. Aim for high-performance, near-zero-net-energy buildings now.
8. Measure, verify, and report performance to improve processes and change behavior.
9. Use performance-based approaches to unleash the creativity of contractors.
10. Collaborate to drive the market for sustainable products and high-performance technologies.
11. Use standards and guidelines to drive change and embed sustainability into decision-making processes.
12. Communicate successes and learn from others.

Systems-based thinking provides a life-cycle perspective that can overcome challenges posed by the federal budget process and by segmented work processes. As importantly, it can help federal agencies identify new ways to use resources, to substitute more sustainable resources, to eliminate waste, and to avoid narrowly focused solutions with unintended consequences.

Systems-based thinking begins with the development of goals and objectives for the activity: The more ambitious the goals, the more innovative the strategies are likely to be. A systems-based approach can be especially effective in helping federal agencies meet their goals for reducing greenhouse gas emissions, reducing the use of potable water, conserving and protecting water resources, for recycling and pollution prevention, for minimizing the generation of waste and pollutants through source reduction, and for regional transportation planning.

Individuals and organizations are using systems-based thinking for reducing greenhouse gas emissions (Oberlin, Ohio; Arlington County, Virginia), for reducing overall energy use and water use (Arlington County, Virginia; Fort Carson, Colorado; Greensburg, Kansas), and for regional transportation planning (Fort Carson, Colorado).

Available and emerging tools to support systems-based thinking include NIST's Building for Environmental and Economic Sustainability (BEES) software, tools developed by the Athena Institute and One Planet Communities, and EARTHSTER. These tools enable the life-cycle assessment of building materials, components, assemblies, and buildings themselves.

2. Focus on community- and regional-based approaches to fill gaps, leverage resources, and optimize results.

Where federal facilities occupy large, contiguous land areas, such as military bases, research campuses, office parks, embassy compounds, and the like, they have opportunities to save energy, reduce the use of fossil fuels, and reduce greenhouse gas emissions by building on-site combined heat and power (co-generation) plants, installing solar arrays and wind turbines for on-site generation of renewable energy (e.g., Oberlin, Ohio; Fort Carson, Colorado; Los Angeles Community College District), and installing district energy systems (Arlington County, Virginia). Larger-scale development also facilitates the recycling of potable water and stormwater management.

Most federal facilities are dependent, in part, on nonfederal infrastructure systems for power, water, wastewater removal, transportation, and telecommunications. Federal agencies can leverage their available resources and achieve goals for strengthening the vitality and livability of adjacent communities by forming partnerships with local communities, utility companies, and others with shared interests (Fort Carson, Colorado). Power purchase agreements can be an effective method to leverage federal land and buildings to achieve energy-saving objectives.

3. Align existing federal facilities to current missions and consolidate the total facilities footprint to lower costs, reduce carbon emissions, reduce water and energy use, and optimize available resources.

Effective portfolio-based facilities management optimizes the performance of existing buildings and other facilities in support of an organization's mission, carefully considers the addition and location of new buildings, and uses life-cycle costing for all potential investments. Federal agencies can use portfolio-based management to align their facilities with mission (e.g., Fort Carson, Colorado; Office of Overseas Buildings Operations of the State Department); to determine which facilities are excess; to identify noncapital solutions for providing required services and avoid the long-term cost and environmental impacts of new buildings (e.g., virtual embassies); to choose sustainable locations for new buildings; to determine where space can be consolidated; and to optimize the performance of existing buildings. Effective portfolio-based facilities management can help agencies meet an array of environmental and cost objectives associated with high-performance facilities.

To effectively implement a portfolio-based facilities management approach, federal agencies need a well-trained workforce. The Federal Buildings Personnel Training Act of 2010, when implemented, should help federal managers strengthen the skills of their workforces for operating high-performance buildings and for portfolio-based facilities management.

Tools that can support portfolio-based management include "Recommendations on Sustainable Siting for Federal Facilities" and EPA's *Portfolio Manager*.

4. Operate facilities efficiently to optimize their performance.

The vast majority of facilities that federal agencies will be using in 2020, 2030, and 2040 exist today. Operating building systems as they were designed can result in significant reductions in the consumption of energy and water and can contribute positively to all aspects of indoor environmental quality.

The performance of building systems can significantly decline over time due to improper installation, the lack of routine maintenance and repair, and simple wear and tear. Such decline means that energy, water, and other resources are wasted, possibly affecting the health and safety of occupants. Building commissioning is a well-recognized best practice for ensuring that building systems operate as they were designed, which can result in lower energy consumption and improved indoor environmental quality. For existing buildings, the recommissioning of heating, ventilation, and air-conditioning (HVAC) systems can be cost effective when undertaken every 3 to 5 years.

5. Aggressively implement proven sustainable technologies as a matter of course.

Agencies regularly replace worn-out roofs, lighting systems, heating, ventilation, and air-conditioning systems, water fixtures, computers, printers, and other equipment in existing buildings. Federal agencies have significant opportunities to upgrade the performance of existing building systems through effective operations, through routine maintenance, repair, and replacement programs, and through retrofit projects. As systems are changed out, more efficient technologies can be incorporated to reduce greenhouse gas emissions and energy and water use, to improve indoor environmental quality, and to meet other objectives related to high-performance green buildings.

Technologies are available that have been proven to reduce greenhouse gas emissions (e.g., white roofs), energy, and/or water use (e.g., Energy Star rated appliances and equipment, WaterSense fixtures, lighting components, FEMP-designated electronics) and that can be incorporated into existing facilities through routine maintenance, repair, and replacement programs. Energy Savings Performance Contracts can be an effective way to improve the efficiency of building systems when agencies lack the up-front capital to directly invest in such improvements.

When more extensive retrofits of building systems are undertaken, agencies can introduce more efficient technologies for heating, lighting, and cooling, for fossil-fuel reduction, and for increased use of renewable energy (e.g., solar panels on the rooftops of buildings, parking garages, and other facilities). Commissioning of retrofitted systems can ensure that they will perform as designed. The addition of new monitoring systems can lead to more efficient operations and can be used as a communication tool to help change behavior.

6. Use integrated, collaborative processes and practices to overcome conventional segmented processes that fail to optimize resources.

Integrated, collaborative work processes are essential for achieving the multiple objectives associated with high-performance buildings, including zero-net-energy buildings, such as NREL's research support facility. They can be used to overcome the wasting of resources inherent in conventional, segmented processes and to support a life-cycle perspective.

Agencies could leverage available resources, meet public policy goals, and improve results now and over the long term by consistently implementing existing guidelines such as those in the "Guiding Principles for Federal Leadership in High Performance and Sustainable Buildings." Even greater reductions of energy use could be achieved if, during the design process, agencies considered the energy required to operate lighting, computers, servers, copy machines, appliances, and other equipment.

Hundreds of tools and models are available to help agencies evaluate alternative designs for buildings and help to optimize the use of natural resources in providing lighting, energy, and other services. Technologies to support integrative design processes, such as building information modeling (BIM), are being developed and used by some agencies for some applications.

7. Aim for high-performance, near-zero-net-energy buildings now.

The technologies and integrated design processes needed to develop high-performance buildings, including near-zero-net-energy buildings, are already available, and some agencies are using them effectively (e.g., NREL research support facility). Federal agencies that wait until 2020 to begin designing zero-net-energy buildings will be missing a significant opportunity to leapfrog ahead to meet their goals and conserve resources. Starting now also provides the opportunity to learn how best to combine technologies and processes to achieve zero-net-energy buildings for a range of climates and locations, and to share that information with other agencies.

Historic buildings present an opportunity to create zero-net-energy buildings. Many historic buildings were originally designed with passive heating and cooling coupled with natural daylighting and ventilation strategies. However, their performance may have been compromised over time through the accretion of mechanical systems and the elimination of original components. By carefully retrofitting and replacing existing systems, some historic structures can become high-performance buildings again (e.g., Wayne Aspinall federal building).

8. Measure, verify, and report performance to improve processes and change behavior.

Achieving all of the objectives associated with federal high-performance facilities requires changes in mindset as much as it does changes in processes. Change within an organization requires leadership and effective communication so that everyone in the organization understands and accepts that the goals and objectives are the right ones to continuously pursue. Because effective operation of building systems is dependent, in part, on the behavior of building occupants, occupants also need to understand how their behavior affects building performance and why proper operation is important to their own health and safety and to their agency's mission. Best-practice organizations have long used performance measurement as a basis for good communication, for changing conventional processes, and for changing human behavior.

Because an array of performance measures have been developed to track progress toward different goals or objectives related to federal high-performance facilities, some measures conflict and create disincentives for sustainable practices. For example, agencies have been directed to (1) reduce their energy use per square foot of space and (2) reduce their total square footage of space. Reducing total square footage of space should, intuitively, also lead to reduced energy use. However, if an agency is successful in reducing its total square footage of space, its energy use per square foot may increase and it will appear that the agency is failing to meet the objectives. This lack of alignment among performance measures undermines the achievement of what should be complementary objectives. New performance measures are being developed to track greenhouse gas emissions and carbon footprint. To the extent possible, the government and its agencies should ensure that all performance measures are aligned to achieve complementary objectives.

Other techniques, technologies, and tools that can be used by agencies to improve communication and to help change behavior in support of the range of objectives associated with high-performance buildings include the following:

- Providing regular updates to stakeholders on progress in achieving objectives related to high-performance buildings through Webinars, conferences, and other formats.
- Using "branding" techniques, including green building rating systems, logos, and energy performance labels for buildings as well as equipment.
- Using evidence-based information for making a "business case" for high-performance buildings and sustainable practices.
- Using energy audits and other reporting requirements to communicate with decision makers, building operators, and occupants about successes and inefficiencies, and about ways to further improve efforts aimed at energy reduction or other goals.
- Using real-time monitoring and feedback systems.
- Establishing friendly competitions among building tenants to boost efforts aimed at reducing energy and water use.

9. Use performance-based approaches to unleash the creativity of contractors.

When new buildings or major retrofits are needed, federal agencies develop criteria for the projects and then contract with private-sector firms to design and construct them. Federal agencies can use performance-based contracts to set high-level performance goals for new buildings and major retrofits and then challenge private-sector contractors to use their creativity and expertise to design projects that meet those goals.

When several years have elapsed between the actual design of a project and its construction, the designs can go stale, such that the project will not be state of the art when the ribbon is cut. In these circumstances, agencies should work with contractors through charrettes or other practices to update the designs to state-of-the-art standards before construction.

The National Renewable Energy Laboratory (NREL) emphasized the use of performance-based contracting strategies to achieve higher-level goals through innovative, creative solutions developed by their contractors.

One potential best practice identified in the workshop breakout sessions was to make reduced operations and maintenance costs an evaluation selection criterion in requests for proposals and contractor selection for new and retrofit projects. In this way, contracts could help incorporate a life-cycle perspective. Another suggestion from the workshop was to develop model clauses and best practice procedures for using performance-based approaches and post them on a public Web site accessible to all federal agencies and their contractors.

10. Collaborate to drive the market for sustainable products and high-performance technologies.

Federal agencies can use their purchasing power to drive the market demand for sustainable products and services, such as was done by the LACCD in developing sustainable carpeting. Realizing such opportunities will require agencies to collaborate with each other and with industry, universities, and nonprofit entities in public-private partnerships. Agencies can also drive the demand for high-performance space through their leasing practices, as recognized in Executive Order 13514. Agencies can help drive the development of technologies such as BIM by providing incentives for contractors to use these technologies for new building and major retrofit projects.

Federal agencies have the opportunity to drive the wider deployment of new, more resource-efficient technologies and products by using their facilities as test beds for new technologies and practices and then publicizing the test results. In this way, agencies and the private sector can create a knowledge base for new technologies and practices that will help to mitigate the risk of using them.

11. Use standards and guidelines to drive change and embed sustainability into decision-making processes.

Federal agencies can meet objectives for high-performance buildings by embedding sustainable practices into their policies, design standards, and acquisition and maintenance practices and through the use of guidelines such as green building rating systems.

Many agencies maintain their own sets of design and operations standards to address the types of buildings they typically manage. One relatively easy way to embed sustainability into everyday decision making is to review these standards and revise them as necessary to align with objectives for high-performance green buildings. Specifying Energy Star appliances and equipment, WaterSense fixtures, and FEMP-designated electronics in contracts and task orders would result in improved energy and water performance almost automatically.

12. Communicate successes and learn from others.

Sustainable practices and processes are evolving and proliferating rapidly. Federal agencies have already developed numerous databases and Web sites containing policies, guidelines, processes, tools, technologies, and evidence-based data for developing, operating, retrofitting, and managing high-performance green buildings and facilities. Among these are *Greening Federal Facilities: An Energy, Environmental, and Economic Resource Guide for Federal Facility Managers and Designers* (DOE, 2001), the *Federal Research and Development Agenda for Net-Zero Energy, High-Performance Green Buildings* (NSTC, 2008), the Whole Building Design Guide (www.wbdg.org), the High-Performance Federal Buildings Web site (http://femp.buildinggreen.com/), Energy Star (www.energystar.gov), and WaterSense (http://www.epa.gov/WaterSense) programs for efficient equipment, appliances, and fixtures, the electronic product environmental assessment tool (www. http://www.epa.gov/epp/pubs/products/epeat.htm), the Building Energy Software Tools Directory (http://apps1.eere.energy.gov/buildings/tools_directory/), and the newly released Sustainable Facilities Tool (http://www.sftool.org/). However, these Web sites and databases are scattered among many individual agencies and their overall value is diminished by this dispersal.

Federal agencies should collaborate to determine how they can best optimize the value of such information so that it can be used more effectively by all federal agencies and so that it can be easily shared with state and local governments, private-sector and not-for-profit organizations, and the public. When agencies test new technologies and practices, they could also place the results on a Web site to help deploy effective technologies to a wider audience.

References

AIA (American Institute of Architects). 2007. Integrated Project Delivery Guide. Washington, D.C.: American Institute of Architects.

Brown, R., S. Borgeson, J. Koomey, and P. Biermayer. 2008. U.S. Building-Sector Energy Efficient Potential. Berkeley, Calif: Lawrence Berkeley National Laboratory.

CIB (International Council for Research and Innovation in Building and Construction). 2009. Conference Report on CIB IDS2009—Improving Construction and Use through Integrated Design Solutions. Available at http://heyblom.websites. xs4all.nl/website/newsletter/0907/ids2009.pdf. Accessed February 28, 2011.

Daily Sentinel. Fed Building Project Aims for 'Net Zero.' Gary Harmon. Thursday, December 2, 2010.

DOE (Department of Energy). 2001. Greening Federal Facilities: An Energy, Environmental, and Economic Resource Guide for Federal Facility Managers and Designers. Second Edition. Washington, D.C.: Department of Energy.

DOE. 2008. Buildings Energy Data Book, Section 1.1.1. Available at http://buildingsdatabook.eren.doe.gov/. Accessed February 28, 2011.

Eichholtz, P., N. Kok, and J. Quigley. 2009. "Doing Well by Doing Good? Green Office Buildings." Working Paper No. W08-001. University of California at Berkeley.

FFC (Federal Facilities Council). 2003. Starting Smart: Key Practices for Developing Scopes of Work for Facilities Projects. Federal Facilities Council Technical Report No. 146. Washington, D.C.: The National Academies Press.

GAO (Government Accountability Office). 2009. Federal Real Property: An Update on High Risk Issues. GAO-09-801T. Washington, D.C.: GAO.

GAO. 2011. Federal Real Property: The Government Faces Challenges to Disposing of Unneeded Buildings. GAO 11-370T. Washington, D.C.: GAO.

GSA (General Services Administration). 2009. Fiscal Year 2008 Federal Real Property Report. Washington, D.C.: GSA. Available at http://www.gsa.gov/graphics/ogp/FY_2008_Real_Property_Report.pdf. Accessed February 28, 2011.

GSA. 2010. Fiscal Year 2009 Federal Real Property Report. Washington, D.C.: GSA. Available at http://www.gsa.gov/graphics/ogp/FY2009_FRPR_Statistics.pdf. Accessed February 28, 2011.

IFMA Foundation (International Facility Management Association Foundation). 2010. Sustainability "How To" Guide Series. Green Building Ratings Systems. Houston, Texas: IFMA. Available at www.ifma.org. Accessed January 10, 2011.

Jones, S. 2009. Introduction to BIM: Smart Market Report Special Section. Pages 21-24 in Building Information Modeling: Transforming Design and Construction to Achieve Greater Industry Productivity. New York, N.Y.: McGraw-Hill Construction.

Kats, G. 2010. Greening Our Built World: Costs, Benefits, and Strategies. Washington, D.C.: Island Press.

Kidd, R. 2010. Statement of Richard Kidd, Program Manager, Federal Energy Management Program, Office of Energy Efficiency and Renewable Energy, U.S. Department of Energy, Before the Subcommittee on Federal Financial Management, Government Information, Federal Services and International Security, Committee on Homeland Security and Government Affairs, U.S. Senate. January 27, 2010. Available at http://hsgac.senate.gov/public/index.cfm?FuseAction=Hearings. Hearing&Hearing_ID=c7cb1779-8aa1-4250-8dfe-18e06b579af1. Accessed February 28, 2011.

Mills, E. 2009. Building Commissioning: A Golden Opportunity for Reducing Energy Costs and Greenhouse Gas Emissions. Berkeley, Calif: Lawrence Berkeley National Laboratory. Available at http://cx.lbl.gov/documents/2009-assessment/LBNL-Cx-Cost-Benefit.pdf. Accessed February 28, 2011.

NAS-NAE-NRC (National Academy of Sciences-National Academy of Engineering-National Research Council). 2008. What You Need to Know About Energy. Washington, D.C.: The National Academies Press.

NAS-NAE-NRC (National Academy of Sciences-National Academy of Engineering-National Research Council). 2010. Real Prospects for Energy Efficiency in the United States. Washington, D.C.: The National Academies Press.

NAS and NAPA (National Academy of Sciences and National Academy of Public Administration). 2010. Choosing the Nation's Fiscal Future. Washington, D.C.: The National Academies Press.

NIST (National Institute of Standards and Technology). 2004. Cost Analysis of Inadequate Interoperability in the U.S. Capital Facilities Industry. NIST GCR 04-867. Gaithersburg, Md.: NIST.

NRC (National Research Council). 1998. Stewardship of Federal Facilities: A Proactive Strategy for Managing the Nation's Public Assets. Washington, D.C.: National Academy Press.

NRC. 2004. Investments in Federal Facilities: Asset Management Strategies for the 21st Century. Washington, D.C.: The National Academies Press.

NRC. 2007. Green Schools: Attributes for Health and Learning. Washington, D.C.: The National Academies Press.

NRC. 2008. Core Competencies for Federal Facilities Asset Management Through 2020: Transformational Strategies. Washington, D.C.: The National Academies Press.

NRC. 2009. Advancing the Competitiveness and Efficiency of the U.S. Construction Industry. Washington, D.C.: The National Academies Press.

NSTC. (National Science and Technology Council). 2008. Federal Research and Development Agenda for Net-Zero Energy, High-Performance Green Buildings. Washington, D.C.: NSTC.

PCAST (President's Council of Advisors on Science and Technology). 2010. Report to the President on Accelerating the Pace of Change in Energy Technologies Through an Integrated Federal Energy Policy. Washington, D.C.: PCAST. Available at http://www.whitehouse.gov/sites/default/files/microsites/ostp/pcast-energy-tech-report.pdf. Accessed February 28, 2011.

Rubinstein, F., and A. Enscoe. 2010. Achieving Energy Savings with Highly-Controlled Lighting in an Open-Plan Office. Berkeley, Calif: Lawrence Berkeley National Laboratory. Available at http://escholarship.org/uc/item/2bt765jw?query=Rubinstein"#page-1. Accessed February 28, 2011.

Appendixes

A

Biosketches of Committee Members

David J. Nash, *Chair,* U.S. Navy, Civil Engineer Corps (retired), is a senior vice president with MELE Associates, Inc., and the president of Dave Nash and Associates, a project development firm serving businesses and governments worldwide. The firm provides project and program management services throughout the world's emerging markets for bioenergy, energy, and large infrastructure projects. RADM Nash was elected to the National Academy of Engineering (NAE) in 2007 for leadership in the reconstruction of devastated areas after conflicts and natural disasters. He is the current chair of the National Research Council's (NRC) Board on Infrastructure and the Constructed Environment (BICE). From 2005 until January 2007, RADM Nash was the president of Government Operations at BE&K, Inc., an international design-build construction firm. In 2003 and 2004, he served as the director of the Iraq Reconstruction Program. He was formerly the president of PB Buildings and manager of the Automotive Division of Parsons Brinckerhoff Construction Services, Inc. RADM Nash completed his 33-year career in the U.S. Navy as the Chief of the Naval Facilities Engineering Command and Chief of Civil Engineers.

RADM Nash is a member of the NAE's Civil Engineering Peer Committee (2009-2012) and the NRC's Committee on National Security Implications of Climate Change on U.S. Naval Forces. He has served as chair of the NRC committee that produced the report *Sustainable Critical Infrastructure Systems: A Framework for Meeting 21st Century Imperatives;* vice chair of the NRC Committee on Business Strategies for Public Capital Investment, which produced the study *Investments in Federal Facilities: Asset Management Strategies for the 21st Century;* and chair of the NRC committee that authored the 2007 report *Core Competencies for Federal Facilities Asset Management Through 2020: Transformational Strategies.* He is a member of the National Academy of Construction, the Society of American Military Engineers, the American Society of Civil Engineers, the National Society of Professional Engineers, the Institute of Electrical and Electronics Engineers, and the American Society of Quality Control.

Robert Berkebile, FAIA (Fellow of the American Institute of Architects), is a founding principal of BNIM Architects and has contributed 44 years to the architectural profession. He is a pioneer in restorative design with the goal of integrating social, environmental, and economic vitality. His sustainable design and planning projects range from redeveloping plans for the former naval base in Charleston, South Carolina, to restoring towns along the Mississippi River severely damaged by natural disasters, including New Orleans. He has participated in a variety of activities including Greening of the White House, Greening of the Pentagon, and Greening of the Grand Canyon.

As the founder of the American Institute of Architects' (AIA's) National Committee on the Environment, Mr. Berkebile has been one of the central forces behind a new focus on sustainable building that has influenced thousands of architects and changed the face of green architecture in America. He is a founding member of the U.S. Green Building Council and there he helped to develop the council's Leadership in Energy and Environmental Design, or LEED, rating system, a voluntary, consensus-based standard for developing high-performance, sustainable buildings. Since its inception in 1998, LEED has grown to encompass more than 14,000 projects in the United States and 30 countries.

In 2009, Mr. Berkebile received a Heinz Award from Teresa Heinz and the Heinz Family Foundation for his role in promoting green building design and for his commitment and action toward restoring social, economic, and environmental vitality to America's communities through sustainable architecture and planning. He holds a degree in architecture from the University of Kansas and is a registered architect in five states.

Hillary Brown, FAIA, is a principal of New Civic Works, a firm which assists public and institutional clients in greening their facility capital programs. As founder of the Office of Sustainable Design with New York City's Department of Design and Construction, she oversaw that office's 1999 collaboration with the Design Trust and the *High Performance Building Guidelines*, and more recently she co-authored the *High Performance Infrastructure Guidelines*. Ms. Brown was managing editor of the nationally and internationally recognized *City of New York High Performance Building Guidelines*, co-author of the U.S. Green Building Council's *State and Local Green Building Toolkit*, and author of *Implementing High Performance Buildings*. Additionally, she envisioned and co-authored the recently released *High Performance Infrastructure: Best Practices for the Public Right-of-Way* for New York City and the Design Trust for Public Space.

Currently a practicing architect at the firm New Civic Works Ms. Brown specializes in green design for schools, universities, public buildings, and infrastructure. Previously having served on the architecture faculties at the Yale, Columbia, and Princeton University Schools of Architecture, today she is a professor of architecture at the City College of New York's (CCNY's) Spitzer School of Architecture. She leads the school's contribution to CCNY's new interdisciplinary master's program: Sustainability in the Urban Environment, given together with the Grove School of Engineering and CCNY's Division of Science.

Ms. Brown has served on the board of directors of the U.S. Green Building Council and is now a board member for the nationally recognized Healthy Schools Network. A graduate of the Yale University School of Architecture, she has been a Loeb Fellow at the Harvard University Graduate School of Design and a Bosch Public Policy Fellow at the American Academy in Berlin, where she examined green building practices in Germany.

Vivian Loftness, FAIA, is an internationally renowned researcher, author, and educator with more than 30 years of focus on environmental design and sustainability, advanced building systems and systems integration, and climate and regionalism in architecture, as well as design for performance in the workplace of the future. Supported by a university-building industry partnership, the Advanced Building Systems Integration Consortium of Carnegie Mellon University, she is a key contributor to the development of the Intelligent Workplace—a living laboratory of commercial building innovations for performance—and has been the author of a range of publications on international advances in the workplace. Her work has influenced both national policy and building projects, including the Adaptable Workplace Lab at the U.S. General Services Administration and the Laboratory for Cognition at Électricité de France.

As a result of her research, teaching, and professional consulting, Ms. Loftness received the 2002 National Educator Honor Award from the American Institute of Architecture Students and a 2003 "Sacred Tree" Award from the U.S. Green Building Council (USGBC). She has bachelors of science and a masters of architecture degrees from the Massachusetts Institute of Technology and is a registered architect. She serves on the National Board of the USGBC, AIA's Committee on the Environment (2005 national chair), and the Department of Energy's Federal Energy Management Advisory Council.

Ms. Loftness has served on numerous committees of the National Research Council including the following: Committee on the Effect of Climate Change on Indoor Air Quality and Public Health, Committee to Review and Assess the Health and Productivity Benefits of Green Schools, Committee to Assess Techniques for Developing Maintenance and Repair Budgets for Federal Facilities, Committee on Advanced Maintenance Concepts for Buildings, Committee on High Technology Systems for Buildings, and the Panel on Climate-Related Data. Ms. Loftness was a member of the Board on Infrastructure and the Constructed Environment (1996-1999). She has delivered three congressional testimonies on sustainable design.

James B. Porter, Jr., is the founder and president of Sustainable Operations Solutions, LLC, which provides consulting services to help companies make significant and sustainable improvements in workplace safety, process safety management, capital effectiveness, and operations productivity. He previously spent 40 years with the DuPont Corporation, from which he retired as chief engineer and vice president-engineering and operations. Mr. Porter joined DuPont in 1966 as a chemical engineer in the Engineering Service Division (ESD) field program at the Engineering Test Center in Newark, Delaware. In his 40 years with DuPont, he served in a number of management positions, including those in the areas of construction, investment engineering, and facilities design. With the restructuring of DuPont Engineering in November 1990, he became director-engineering operations and was subsequently named director of operations for the Fluoroproducts business (1992), director of operations (1995), vice chair of the DuPont Corporate Operations Network (1995), vice president of engineering (1996), and vice president of safety, health, and environment and engineering (2004). He is currently consulting with several companies on various aspects of construction and facilities management.

Mr. Porter is a member of the National Research Council's Board on Infrastructure and the Constructed Environment. He has served as chair of the Construction Industry Institute (CII) and Delaware's United Negro College Fund. He was the 2004 recipient of CII's Carroll H. Dunn Award of Excellence and in 2005 received the Engineering and Construction Contracting Association Achievement Award. He is a member of the board of governors for the Argonne National Laboratory, the board of directors for the American Institute of Chemical Engineers (AIChE), FIATECH, the Mascaro Sustainability Initiative, and the Fieldbus Foundation. He also participates on various industry advisory boards, including AIChE's Center for Chemical Process Safety. Mr. Porter is a member of the University of Tennessee's College of Engineering Board of Advisors and the National Academy of Construction. He holds a B.S. degree in chemical engineering from the University of Tennessee.

Harry G. Robinson III is a design professional educated in architecture, city planning, and urban design and is currently professor of urban design and dean emeritus and advisor to the president of Howard University and principal of TRG Consulting, an international design firm.

During the period 1979-1995, he served as dean and professor of urban design, School of Architecture and Planning, Howard University, and in 1995-1999 as interim vice president of academic affairs and vice president for university administration, Howard University. Prior to the decanal appointment at Howard University in 1979, Mr. Robinson was the director of the Center for Built Environment Studies that he founded at Morgan State University. This set of programs—architecture, city planning, landscape architecture, and urban design—established that university's first professional interdisciplinary curricula.

Mr. Robinson is a twice presidentially appointed commissioner and elected chair, United States Commission of Fine Arts, and he was elected president of two national architectural organizations: the National Architectural Accrediting Board, 1996, and the National Council of Architectural Registration Boards, 1992. He chaired the UNESCO International Commission on the Goree Memorial and Museum that was established to guide the development of this project in Dakar, Senegal. He has served on major boards and commissions, including the Vietnam Veterans Memorial Fund, the Committee for the Preservation of the White House, the White House Historical Association, and the Kennedy Center for the Performing Arts.

The recipient of the Tau Sigma Delta Architectural Honor Society Silver Medal, Mr. Robinson has been elected to membership in the American Institute of Architects' College of Fellows and honorary membership in the Colegio de Arquitectos de Mexico, Sociedad de Arquitectos Mexicanos, and in the Trinidad and Tobago Institute of Architects. In 1999 he was awarded the Richard T. Ely Distinguished International Educator Award by the Lambda Alpha International Honorary Land Economics Society. In 1991, he had a partial Fulbright Fellowship at the Cooperbelt University, Kitwe, Zambia. Mr. Robinson's awards from the National Organization of Minority Architects include an honor award in 1991 and a special award in 1992. In 1993 Hampton University awarded him its 125th Anniversary Citation for Leadership in Architecture.

Mr. Robinson holds professional degrees in architecture and city planning, B.Arch. with design honors, and MCP, Howard University, and an advanced degree in urban design, MCP in urban design, Harvard University Graduate School of Design.

Arthur H. Rosenfeld is a professor of physics (emeritus) at the Lawrence Berkeley National Laboratory (LBNL) and a member of the California Energy Commission. Dr. Rosenfeld was elected to the National Academy of Engineering in 2010 "for leadership in energy efficiency research, development, and technology deployment through the development of appliance and building standards and public policy." After completing his graduate studies, Dr. Rosenfeld went to the University of California, Berkeley, where he joined, and eventually led, the Nobel prize-winning particle physics group of Luis Alvarez at LBNL, until 1974. At that time, he changed to the new field of efficient use of energy, formed the Center for Building Science at LBNL and led it until 1994. The center developed electronic ballasts for fluorescent lamps (which led to compact fluorescent lamps), low-emissivity windows, and the DOE-2 computer program for the energy analysis and design of buildings. He received the Szilard Award for Physics in the Public Interest in 1986 and the Carnot Award for Energy Efficiency from the U.S. Department of Energy (DOE) in 1993. In 2006, Dr. Rosenfeld received the Enrico Fermi Award, the oldest and one of the most prestigious science and technology awards given by the U.S. government. Dr. Rosenfeld is a co-founder of the American Council for an Energy Efficient Economy, the University of California's Institute for Energy Efficiency, and the Washington-based Center for Energy and Climate Solutions. From 1994 to 1999, Dr. Rosenfeld served as senior adviser to the DOE's assistant secretary for energy efficiency and renewable energy. He received a Ph.D. degree in physics from the University of Chicago.

E. Sarah Slaughter is the associate director for buildings and infrastructure in the Massachusetts Institute of Technology (MIT) Energy Initiative. Her current research focuses on innovations for sustainable and disaster-resilient infrastructure and the built environment. Previously she was co-founder and the head of the Sustainability Initiative in the MIT Sloan School of Management, focusing on strategies for sustainable organizations and communities. From 1999 through 2006, Dr. Slaughter founded and ran MOCA Systems, Inc., a technology firm that developed a construction simulation software system. Prior to establishing MOCA Systems, Dr. Slaughter was an assistant professor in the Department of Civil and Environmental Engineering at MIT, where her research and teaching interests focused on construction management and engineering, innovation in building and infrastructure systems, and computer-aided process simulation of construction activities. Earlier, she was a professor of civil and environmental engineering at Lehigh University and conducted research in the National Science Foundation Center for the Advancement of Large Structural Systems. Dr. Slaughter was named a National Academy Associate for her service on the National Research Council (NRC) Panel on Building and Fire Research, the Committee on Outsourcing Design and Construction Management Services for Federal Facilities, the Committee on Infrastructure Technology Research Agenda, and the Board on Infrastructure and the Constructed Environment (1998-2001; 2007-2011).

Dr. Slaughter is currently a member of the NRC Standing Committee on Defense Materials, Manufacturing, and Infrastructure. She is also a member of the Massachusetts Sustainable Water Management Advisory Board and of the Sustainability Committee in the International Facilities Management Association (IFMA), and she serves on several national advisory committees and editorial boards of professional publications. Dr. Slaughter has published more than 50 articles and books and is a recognized expert in the field of sustainable facility assets and in innovations in the built environment. She received her S.B. in civil engineering and anthropology, S.M. in civil engineering and technology policy, and a multidisciplinary Ph.D. degree in civil engineering and management science from the Massachusetts Institute of Technology.

Clyde B. (Bob) Tatum is the Obayashi Professor of Engineering at Stanford University. He joined the Stanford construction faculty in 1983 after having had nearly 15 years of experience in heavy industrial and military construction. He served as the coordinator of the construction program from 1996 to 1999 and became the department chair in 1999. He is a mechanical engineering graduate of the Virginia Polytechnic Institute (B.S.M.E. 1966) and the University of Michigan (M.S.E. 1970), and he earned a Master of Business Administration from New York University. Dr. Tatum has taught courses on construction engineering and mechanical and electrical systems for buildings in Stanford's graduate construction program and undergraduate civil engineering curriculum, high-tech and industrial construction, concrete construction, management of technology, case studies in managing construction projects, cost engineering, and materials management. His industry experience included responsibility as a mechanical engineer, construction engineer, resident engineer, and construction superintendent/ area manager with Ebasco Services Incorporated (1970-1981) on two large power plant projects. He is a registered professional engineer in Colorado and Washington. In 1986 he received the Presidential Young Investigator Award from the National Science Foundation, and in 1988 he received the American Society of Civil Engineers' Construction Management Award. He was elected to the National Academy of Construction in 2002. He recently served on the National Research Council Committee to Evaluate Future Strategic and Energy Efficient Alternatives for the Delivery of Utility Services to the U.S. Capitol Complex.

B

Committee Meetings and Speakers

JUNE 17-18, 2010

Kevin Kampschroer, Director, Office of Federal High-Performance Green Buildings, U.S. General Services Administration

Michele Moore, Director, Office of the Federal Environmental Executive

Shyam Sunder, Director, Engineering Laboratory, National Institute of Standards and Technology, and Co-Chair, Building Technology Research and Development Subcommittee of National Science and Technology Council

JULY 20-21, 2010

Kevin Kampschroer, Director, Office of Federal High-Performance Green Buildings, U.S. General Services Administration

Shyam Sunder, Director, Engineering Laboratory, National Institute of Standards and Technology, and co-chair, Building Technology Research and Development Subcommittee of National Science and Technology Council

Greg Kats, President of Capital-E and Venture Partner at Good Energies

Peter Garforth, President, Garforth International, LLC

Robert Berkebile, Principal, BNIM Architects

Jeffrey M. Baker, Director of Lab Operations, National Renewable Energy Laboratory, Department of Energy

Christopher Juniper, Sustainability Planner, Fort Carson, Colorado

Hal Alguire, Director of Public Works, Fort Carson, Colorado

Thomas Hall, Facilities Program Manager, Los Angeles Community College District

Roland Risser, Program Manager, Building Technologies Program, Department of Energy, and Co-Chair, Building Technology Research and Development Subcommittee of National Science and Technology Council

NOVEMBER 2-3, 2010

William Miner, Director, Office of Design and Engineering, Office of Overseas Buildings Operations, U.S. Department of State

Katherine "Joni" Teter, Sustainability Subject Matter Expert, Office of Federal High-Performance Green Buildings, U.S. General Services Administration

Arthur Rosenfeld, Professor Emeritus, Lawrence Berkeley National Laboratory

Gregory Norris, Professor, School of Public Health, Harvard University

David Orr, Professor, Environmental Studies Program, Oberlin College

Mark Mykleby, Office of the Chairman, Joint Chiefs of Staff, U.S. Department of Defense

C

Workshop Agenda and List of Participants

FEDERAL HIGH-PERFORMANCE GREEN FACILITIES WORKSHOP
WALTER E. WASHINGTON CONVENTION CENTER
WASHINGTON, D.C.

Agenda

Tuesday, July 20, 2010

1:00 p.m.   Welcome and Opening Remarks. National Research Council Study, Co-Sponsors, Process, and Deliverables
David J. Nash, Chair, NRC Committee on Federal High-Performance Green Buildings

1:15 p.m.   Workshop Objectives and Format
Shyam Sunder, Co-Chair, Building Technology Research and Development Subcommittee of the National Science and Technology Council, and Director, Engineering Laboratory, National Institute of Standards and Technology

1:30 p.m.   Beyond Performance Requirements: Breakthrough Thinking in the Federal Sector
Kevin Kampschroer, Office of Federal High-Performance Green Buildings, General Services Administration

1:45 p.m.   Guest Presentations

The Economics of Sustainability: The Business Case That Makes Itself
Greg Kats, President, Capital-E, and Venture Partner, Good Energies

Beyond Incrementalism: The Case of Arlington, Virginia
Peter Garforth, President, Garforth International, LLC

Transformative Action Through Systems-Based Thinking
Robert Berkebile, Principal, BNIM Architects

2:45 p.m.   Breakouts—Who, What, Where
Shyam Sunder

3:15 p.m.   Concurrent Breakout Sessions

Breakout Topic 1: Investing in and Budgeting for Sustainable Facilities
Breakout Topic 2: Planning, Siting, Infrastructure, and Community Relations
Breakout Topic 3: Sustainable Operations and Maintenance
Breakout Topic 4: Life-Cycle Assessment

4:30 p.m.   Report-outs from Breakout Session Facilitators

Wednesday July 21, 2010

8:30 a.m.   Welcome Back
David J. Nash, Chair, NRC Committee on Federal High-Performance Green Buildings

8:45 a.m.   Guest Presentations

Getting to Net-Zero Energy: NREL's Research Support Facility
Jeffrey M. Baker, Director of Laboratory Operations, National Renewable Energy Laboratory, Department of Energy (DOE)

Sustainable Fort Carson: An Integrated Approach
Christopher Juniper, Sustainability Planner and Hal Alguire, Division of Public Works, Fort Carson, Colorado

Sustainable Asset Management: The Case of Los Angeles Community College District
Thomas L. Hall, Facilities Program Manager, Los Angeles Community College District

Overcoming Regulatory Barriers: What Worked in California
Roland Risser, Co-chair, Building Technology Research and Development Subcommittee of the National Science and Technology Council, and Program Manager, Building Technologies Program, DOE

10:30 a.m.   Concurrent Breakout Sessions

Breakout Topic 5: Design and Construction for New Buildings and Retrofits
Breakout Topic 6: Asset/Program Management
Breakout Topic 7: Sustainable Acquisition
Breakout Topic 8: Regulatory Issues, Voluntary Standards, and Rating Systems

11:45 a.m.   Report-outs from Breakout Session Facilitators

12:30 p.m.   Thank You and Adjournment

List of Participants

Mark Ames, American Society of Heating, Refrigerating and Air-Conditioning Engineers (ASHRAE)
Carolyn Austin-Diggs, General Services Administration (GSA)
Dave Baker, U.S. Department of State (DOS)
Jim Balocki, U.S. Army Corps of Engineers (USACE)
Peter Bardaglio, Second Nature
Michael Bloom, GSA
Catherine Broad, U.S. Department of Agriculture
Corey Buffo, U.S. Department of Energy (DOE)
Lane Burt, Natural Resources Defense Council (NRDC)
Steven Bushby, National Institute of Standards and Technology (NIST)
Philip Columbus, Department of the Army
Anne Crawley, DOE
Jose Cuzme, Indian Health Services
Victor D'Amato, Tetra Tech, Inc.
Lance Davis, GSA
Maria de Isasi, Smithsonian Institution
Ryan Doerfler, GSA
Paul Domich, National Science and Technology Council
Bill Dowd, National Capital Planning Commission (NCPC)
Michael Dunn, U.S. Environmental Protection Agency (EPA)
Terrel Emmons, Office of the Architect of the Capitol (AOC)
Ecton English, National Security Agency (NSA)
Stella Fiotes, NIST
Anna Franz, AOC
Chris Garvin, Terrapin Bright Green
Frank Giblin, GSA
Brad Gustafson, DOE
Ryan Guyer, DOS
Jeffrey Harris, Alliance to Save Energy (ASE)
Byron Haselden, Haselden Construction
Jonathan Herz, GSA
William Holley, GSA
Diana Horvat, Envision Design
Mary Ellen Hynes, Department of Homeland Security
Alison Kinn Bennett, EPA
Bob Kollm, U.S. Postal Service
William Logan, Headquarters, U.S. Coast Guard
Juan Lopez, Office of the Federal Environmental Executive
Philip Macey, Haselden Construction
Sina Mostaghimi, EPA
Get Moy, AECOM Inc.
Steve Pranger, USACE
Douglas Read, ASHRAE
Ab Ream, DOE
Eleni Reed, GSA

Jeffrey Rutt, NSA
Sarah Ryker, Science and Technology Policy Institute
Ken Sandler, GSA
Martin (Marty) Savoie, USACE
Robert Scinta, U.S. Department of Commerce
Graziella Siciliano, ASE
Josh Silverman, DOE
Rodney Sobin, ASE
Diane Stewart, Department of Health and Human Services
Diane Sullivan, NCPC
Shyam Sunder, NIST
Alison Taylor, Siemens Corporation
Joni Teter, GSA
Mary Tidlow, National Park Service (NPS)
Meg Waltner, NRDC
Stephen Whitesell, NPS
David Zimmerman, Tennessee Valley Authority

D

Transformative Action Through Systems-Based Thinking

Robert Berkebile, Fellow of the American Institute of Architects, BNIM Architects

Today I want to share some thoughts and information with you about the tools being used to create sustainable buildings and also about the role of systems-based thinking. While we would benefit from improvements in the technology, tools, and materials needed for sustainable development, the primary limitation is our thinking. So, I'm going to focus on utilizing the available tools and, as importantly, on systematic processes for using those tools to their maximum advantage.

Between 1970 and today, the focus of architecture and construction has evolved. In the 1970s, the focus was on conservation at the building scale. As we began looking at larger and larger scales—neighborhood, city, region, watershed, airshed, jobshed, and so on—we began thinking in terms of sustainability, "Living Buildings," restoration, and regenerative approaches.

From my perspective, systems-based thinking for the U.S. construction industry really began in 1989 through creation of the American Institute of Architects' Committee on the Environment (AIA COTE). The AIA COTE's dialogue with diverse industry and environmental stakeholders gave rise to the U.S. Green Building Council (USGBC).[1] Since then, we have seen dramatic savings in energy, water, and materials from buildings that are certified under the USGBC's Leadership in Energy and Environmental Design (LEED) rating system. Even more importantly, we have seen other benefits, including human health and productivity effects and positive social, economic, and environmental impacts on the surrounding community, which are a direct result of using a more systems-based approach.

The definition of what constitutes good design is evolving. It is a given that building design must satisfy the owner's programmatic needs, including budgets for time and cost, and designs must comply with public safety requirements defined by building codes. But, historically, good design has been a beauty contest. Approximately 25 years ago that definition was called into question when a series of "well-designed," award-winning buildings caused their occupants to be uncomfortable or sick, had excessive operating and maintenance costs, and resulted in negative impacts on the surrounding neighborhood or environment. Today, if a building is not healthy for its occupants, for the planet, and for the future of all life, it is not well designed, no matter how good it looks.

---

[1]Information available at http://www.usgbc.org.

FIGURE D.1 Internal Revenue Service building, Kansas City, Missouri. SOURCE: Courtesy of BNIM Architects.

The "Top Ten Green" building design awards that the AIA COTE bestows every year provide a snapshot of this evolution. On the AIA Web site (http://www.aia.org), one can see how the industry has progressed in its concept of what is green and what is good design.

One recent example of good design recognized in AIA's Top Ten program is a large Internal Revenue Service (IRS) center in my hometown, Kansas City, Missouri (Figure D.1). This building of approximately 1 million square feet of space was constructed using a design-build process in which the IRS, the General Services Administration (GSA), the developer, the city, and the source of financing were all partners.

The design team used building information modeling (BIM) to analyze many scenarios quickly to advance the best concepts—those that would satisfy a range of stakeholders and create an environment within which the people processing tax returns could feel good about their environment and experience increased health and productivity.

BNIM Architects has used BIM technology on a variety of projects including a band shell in North Charleston, South Carolina (Figure D.2). This facility was part of a 3,000-acre redevelopment, and our client wanted it to be ready for a Fourth of July concert. However, by the time design approval was received from the city, we did not have time for a typical design-bid-build process. So, our office e-mailed our design (using BIM) to the contractor. The design documents were created and approved electronically and then entered directly into the contractor's manufacturing system (computer-controlled fabrication). The only piece of paper that was printed for this project was the foundation drawing for the local contractor. Technology and collaboration, in this case, made the impossible possible.

We find that high-performance goals require systems-based thinking, collaboration, and computer tools to facilitate a collaborative dialogue of discovery. A comprehensive redevelopment plan for 3,000 acres in North Charleston, South Carolina, is a good example. We first looked back 12,000 years at the deep ecology of the place to understand its history—not to restore it, but to create the best options for moving forward and adding vitality with each decision. As seen in the conceptual plan in Figure D.3, we were examining human systems of 5- and 10-minute walking circles, centering them on the 11 schools within the area. This approach influenced transportation systems and the co-location of community centers, libraries, health clinics, and other community services within the schools. These decisions, in turn, increased the efficiency, quality of life, and economic performance of our community.

FIGURE D.2 Point Pavilion in North Charleston, South Carolina, designed and delivered using building information modeling. SOURCE: Courtesy of BNIM Architects.

FIGURE D.3 Conceptual plan for North Charleston, South Carolina, using 5- and 10-minute walking circles centered on existing schools. SOURCE: Courtesy of BNIM Architects.

As far as I know, this was the first time in the United States that a developer had entered into an agreement with a city making a commitment that every decision related to design, planning, investment, and construction would increase the vitality of the social, economic, and environmental systems simultaneously. This commitment required holistic, systems-based solutions. The developer and design team could not forward recommendations that included trade-offs between social and/or environmental vitality and economic performance; the recommendations had to improve all three sectors from the community developer's perspective.

This vision was created through collaboration during a 2-year planning process involving a wide range of stakeholders, including community interests, government agencies, neighborhoods, 65 consulting firms, universities, developers, economic consultants, and nonprofit foundations. All stakeholders made decisions in a series of meetings at various venues. Over time, the BNIM design team built a tool to inform and document these decisions, goals, and metrics. We named it the Noisette Rose, after the French botanist who created a beautiful white rose in Charleston (Figure D.4). Sadly his rose was exported to France; none remain in South Carolina, and the stream named for him has been destroyed by urban development and industrial pollution.

The design team built the Noisette Rose tool to inform our team about the design commitments, to inform our client about their investment options, and to provide the community with a way to monitor the results. It tracks the social, environmental, and economic goals; at some point we renamed these "people, planet, and prosperity," which we refer to as the Triple Bottom Line. Each spoke on the graphic is a specific goal, and each goal has a specific metric against which it will be measured. The spoke at three o'clock for example, is energy efficiency. On the project represented here, we exceeded the goal slightly, and so the rose projects beyond the ring. By glancing at this diagram, anyone in the community can see that the project is going well. If the shape is a rosebud, it indicates poor performance. If the shape is asymmetrical, it indicates that some sector needs more attention. With this type of approach,

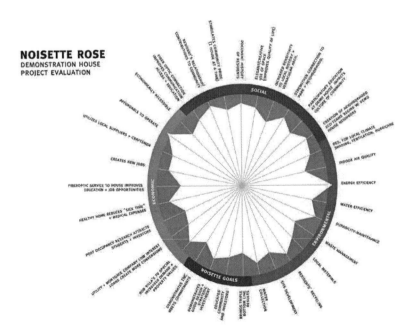

FIGURE D.4 Noisette Rose evaluation tool. SOURCE: Noisette Sustainable Master Plan. Courtesy of BNIM Architects.

FIGURE D.5 Greensburg, Kansas, after the tornado in July 2007. SOURCE: Courtesy of Larry Schwarm.

using these tools and what I call a "creative dialogue of discovery," it becomes much easier to improve the relationships between people, planet, and prosperity.

In 2007, a category EF5 tornado destroyed every structure except the grain elevator in the town of Greensburg, Kansas (Figure D.5). Possibly the second greatest shock to the community, however, was waking up to the headline in the *New York Times,* "Nature Performed a Coup de Grace on Kansas Town" (July 24, 2007). The residents began meeting in tents and other makeshift venues to determine what future they wanted to create. It was dialogue and systems-based thinking that allowed them, in a series of meetings held in the months following the tornado, to develop a unique, high-performance vision for the future of their community (Figure D.6).

Their vision was: "Blessed with a unique opportunity to create a strong community devoted to family, fostering business, and working together for future generations." To implement that vision, they wanted to promote a high level of efficiency in new construction and to look to renewable options for energy generation.

FIGURE D.6 Community meeting in Greensburg, Kansas, after the 2007 tornado. SOURCE: Courtesy of BNIM Architects.

Although the vision was beautiful, it was too vague to provide a basis for design and investment decisions. So the BNIM planning and design team helped the residents of Greensburg develop a set of specific goals related to community, family, prosperity, environment, affordability, growth, renewal, water, health, energy, wind, and the built environment. For example, each goal has specific metrics that inform every project that occurs in Greensburg. These goals and metrics dramatically transformed the master plan and changed the town forever with regard to its buildings' efficiency, performance, and operating costs. There is more connectivity, more biodiversity, and more intimacy with the landscape (Figure D.7) For example, the plan treats every drop of water as a precious resource. Water is captured from buildings and landscape (including the streetscape), purified, stored, used, and returned to groundwater as clean as when it fell from the sky (Figure D.8).

FIGURE D.7 Open space and green corridors system, Greensburg, Kansas. SOURCE: Courtesy of BNIM Architects.

FIGURE D.8 Conceptual design for streetscape, water capture, and storage, Greensburg, Kansas. SOURCE: Courtesy of BNIM Architects.

New city-owned buildings in Greensburg must be twice as efficient as the ones that they replaced and must provide for better health, more comfort, and more passive survivability. Moreover, the city's wind farm will generate far more energy than the city needs, and the excess will be sold to the grid, creating a revenue stream for the city. Greensburg was the first U.S. city to adopt a resolution making USGBC's LEED Platinum its standard building goal. The six buildings completed to date—city hall, a community center, a K-12 school, a business incubator, a regional hospital, and a John Deere dealership—all meet that goal (three of these buildings are shown in Figure D.9).

One year following the tornado, the *New York Times* said, in essence: This is the most brilliant recovery in America. This should be the model for how to build after disaster. My hope is that this approach becomes the model for revitalizing rural America: using systems-based thinking to inform community designs and decisions, including buildings, economy, and lifestyles.

Greensburg is one response to Einstein's statement that "we shall require a substantially new manner of thinking if mankind is to survive." But even though they utilized LEED Platinum—and I would argue that the LEED rating system has been the most transformative tool in the design and construction industry—even LEED Platinum is only third-party certification that one is doing less damage to the environment than everyone else. Surely it is time to move beyond the concept of doing less damage, to doing something positive (restorative or regenerative).

I have been working on this idea since the mid-1990s, when my firm was working on a demonstration project for the National Institute of Standards and Technology (NIST) for the LEED 1.0 rating system. At that time, we called the rating system "Plus Ultra" (Latin for "more beyond"). It evolved with input from many, including Janine Benyus, Jason McLennan, and a major study funded by the David and Lucile Packard Foundation, to become what is known as a "Living Building." The Packard Foundation hired my firm to design a new headquarters, and we signed a contract to deliver a LEED Platinum building. When we shared the idea of moving beyond LEED Platinum (less bad) to something positive, like a Living Building, they commissioned us to create six building designs on their site: a market rate building, the four levels of LEED, and a Living Building.[2] The foundation wanted a comprehensive analysis of the relative costs, timing, and benefits of each level of performance, including 30, 60, and 100 years. When the study was completed, the foundation's chief financial officer said that the only responsible decision, financially, was to design and build a Living Building. Since then, the concept has continued to evolve under the shepherding of the Cascadia chapter of USGBC to become the Living Building Challenge,[3] which we introduced at Greenbuild[4] in 2007.

Living Buildings gain that stature by being audited after their first year of operation to verify that they perform at the level at which they were designed, including generating more energy than they consume and purifying more water than they pollute. A facility that BNIM designed in Rhinebeck, New York, for the Omega Institute was recently the first LEED Platinum building in the world to become a certified Living Building (Figure D.10). It's also the first sewage treatment facility (biological waste water treatment system) that has been claimed as the venue for the institute's yoga classes.

Living Buildings are informed by and heavily rooted in the indigenous characteristics of a building's eco-region in order to renewably generate their own energy; capture, treat, and use their own water; and operate by embracing the essence of what the site can provide. It is very simple but very demanding. There are five typologies within the Living Building Challenge: renovation, building, neighborhood,

---

[2]A Living Building harvests all of its own energy and water, is adapted to the climate of the site, operates pollution-free, promotes health and well-being, is composed of integrated systems, and is beautiful.

[3]For information, see http://ilbi.org/the-standard/lbc-v1.3.pdf.

[4]See http://www.greenbuildexpo.org/Home.aspx.

FIGURE D.9 LEED Platinum buildings in Greensburg, Kansas. NOTE: LEED, Leadership in Energy and Environmental Design, the rating system of the U.S. Green Building Council. SOURCE: Courtesy of BNIM Architects.

FIGURE D.10 Omega Institute for Holistic Studies facility in Rhinebeck, New York. SOURCE: Courtesy of BNIM Architects.

landscape, and infrastructure. Both LEED and the Living Building Challenge started at the building scale but are now moving to the larger, community scale.

Our firm, with a stellar team of consultants that includes Vivian Loftness, is working on a Living Building at the neighborhood scale at the University of Georgia, Athens—the new Odum School of Ecology. Eugene Odum was arguably the father of systems-based thinking in the United States. It seems appropriate that the school named after him would have these attributes and could measurably transform the campus, the city, and the state in terms of utilizing systems-based thinking to achieve new levels of performance. For example, our design for the landscape surrounding the building was created in collaboration with the faculty and nine programs that are part of the curriculum and research. The faculty and administration are helping us articulate priorities that are now being assigned metrics as part of a strategy for achieving their goals (Figure D.11).

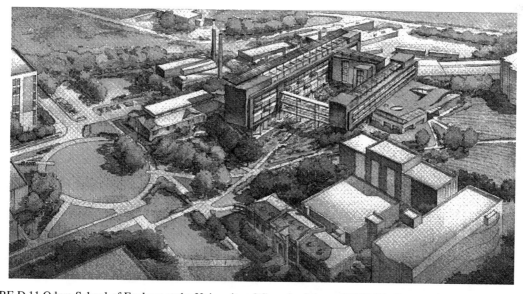

FIGURE D.11 Odum School of Ecology at the University of Georgia, Athens. SOURCE: Courtesy of BNIM Architects.

Over time, BNIM has learned that if we use a systems-based approach it changes the architecture and the landscape, and the line between the two becomes blurred. Often it becomes difficult to distinguish where one ends and the other begins. I find it helpful to try to understand the metabolism of the system. By examining how the whole system operates, we can design ways to close loops or make new connections to add vitality. For example, we like to convert "waste" to a resource by finding ways to put it back into the system as something productive and useful, much like how a natural system would manage waste.

Part of the design process for this project and others is to look for opportunities to create biomimetic materials that improve performance. The first famous biomimetic material used was for the swimsuits that U.S. athletes wore at the Beijing Olympics in 2008, when they broke so many swimming records. The swimsuits were made from a biomimetic fabric that was inspired by analyzing shark skin and dolphin skin. Scientists discovered that the skin of those animals is not smooth, but actually consists of a series of dimples and bumps that reduce friction. That knowledge was used to create the fabrics that dramatically improved the times registered by U.S. swimmers. We believe that incorporating biomimetic materials in buildings could result in building materials that are self-cleaning and self-healing or materials whose thermal value changes as needed.

By utilizing BIM on the Odum building project, or more specifically a green BIM, we can "test-drive" this building and compare it to the two best-performing buildings on campus. (It is not a one-to-one comparison, because one is a school of art and one is a classroom building, but we found that the Odum building out-performs the others, even though laboratories typically use far more resources [energy and water] than are used by other types of buildings.)

We did not plan to analyze carbon offsets, but potential donors were interested in tracking economic performance over time, including a measure for carbon offsets or carbon credits.

Now, I'm going to shift gears. We had hoped that Gregory Norris of the Harvard School of Public Health would be here to talk about life-cycle assessment and transforming the supply chain as it relates to buildings. Because he was unable to come, I was asked to fold some of that information into this presentation.

Work on a life-cycle assessment of the supply chain began in 1993 as part of a research project at Montana State University in Bozeman that was funded by the National Institute of Standards and Technology. The project was a research laboratory, and the subject of the grant was to create a building design that was more energy efficient than any building of its type. While we were interested in energy efficiency, it seemed that we knew a lot more about energy efficiency in 1993 than we knew about human health and productivity, about increases in biodiversity, and about the consumption of water and other materials and resources. I asked if we should not also be looking at those issues.

NIST expanded our commission to look at those issues, and a series of new approaches and tools were created as a result. This was at the time that the USGBC was born, and the building was being designed by many of the same people who were involved in creating the LEED rating system. We explored the possibility of resourcing all materials from within a 500-mile radius, which soon thereafter influenced the LEED rating system. We also looked at converting waste streams to new building materials. In order to inform the decisions, we had to know what the material flows were, although at the time this information was not available; so, we physically located, tracked, and modeled the resources that were available in the region. This research created several interesting new materials, but it was a labor-intensive process.

Several years later, with LEED in place, BNIM began working on a new School of Nursing for the University of Texas Health Science Center at Houston. We knew that we needed a more efficient process to analyze the best material choices, and Gregory Norris agreed to create software that would allow

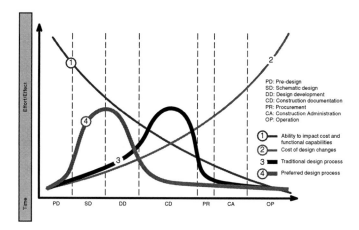

FIGURE D.12 Cost influence curves and Architecture, Engineering, Construction (AEC) productivity. SOURCE: Construction Users' Roundtable.

us to use a large body of data collected on a county-by-county basis by the Environmental Protection Agency (EPA) and the Department of Commerce. The software allowed us to use those data to evaluate the upstream environmental impact of our design decisions. Using this tool on the School of Nursing to gain access to a much larger body of information improved the environmental impacts of our selections, as well as the performance of the building.

But our client was even more excited about the economic impacts of these decisions. An analysis comparing the base case to the final design showed that we improved the economy in Harris County, Texas, by $1.1 million through intentional design decisions.

The good news is that Greg Norris continues to improve this software. Walmart is now using the new generation of this "open source" tool, currently called "EARTHSTER," to communicate with and improve the environmental performance of all of its suppliers. The suppliers and manufacturers can log on, describe their process, and answer a series of questions, and the tool will provide an evaluation of their environmental performance. Assuming a good evaluation, they can send it back to Walmart and qualify to be a supplier. A supplier that does not like an evaluation does not have to share it with Walmart, but EARTHSTER captures all the data. As a result, this tool will generate an open-source, Web-based database on materials.

Fortunately there is a growing family of tools and processes that one may want to consider in order to increase quality and performance or to decrease time and costs. One promising approach is called "lean construction," which changes the relationship between the project owner, designers, and contractors by contract and stimulates a healthy dialogue and partnership among all the stakeholders. One of the things we know is that our ability to have meaningful impact on cost or quality is reduced as the project advances (Figure D.12). If quality information is available early in the project, effective decisions can be made as a collaborative team, improving the outcome dramatically.

Another process that can be utilized if one is working at the community scale (for example, on a military complex, hospital campus, or neighborhood development) is One Planet Communities.[5] This is an excellent program that incorporates a systems-based approach. There are 10 areas of study (Figure D.13).

For each of these areas, the user must establish a specific goal for his or her project. Software is

---

[5] Information available at http://www.oneplanetcommunities.org/about-2/.

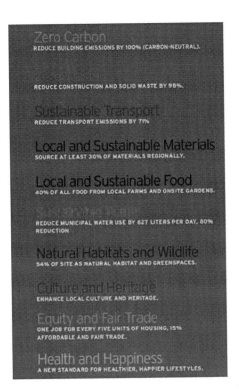

FIGURE D.13 One Planet Communities areas of study. SOURCE: Petite Rivière Regenerative Plan, April 2009.

available to help you assess possibilities during the design-alternatives period of planning, and then, once the project is complete or begins implementation, the user can receive a regular, ongoing follow-up.

One Planet Communities was originally developed by BioRegional (a not-for-profit in the United Kingdom). It was trying to create a prototype community to reduce its environmental footprint in England. At the time BioRegional calculated that if everyone lived like an Englishman, it would require three planets to provide the resources. The goal of the first project, BedZED (Beddington Zero Energy Development), was to provide quality living but reduce consumption to the equivalent of one planet. After measuring the results at BedZED, BioRegional discovered that it had fallen short of its goal and was operating at the equivalent of one and one half planets. It also realized that if resource reduction was its goal, it should be working in North America, where resource consumption is closer to the equivalent of five to six planets. So, BNIM is working on a redevelopment project in Montreal, which is a golf course surrounded by existing development and next to the rail. Over time, it will be transformed into a community with more biodiversity than existed when it was a golf course and hopefully will serve as a model for living successfully in the 21st century.

One Planet Communities, like the Living Building Challenge, is simple. Both require a shift in thinking clarity and a willingness to embrace very high goals. As Kevin Kampschroer said earlier during this workshop, "It's about claiming the future and then living into it." Buckminster Fuller taught me that the best way to predict the future is to design it. He also believed that we are all born geniuses, and that we are gradually "de-geniused" by our parents and our teachers. I believe these initiatives come at the perfect time for us to reclaim our genius—by improving the quality of our dialogue with better tools, better information, and inspiring one another to create 21st-century regenerative solutions.

E

Sustainable Fort Carson: An Integrated Approach

Christopher Juniper, Sustainability Planner, and
Hal Alguire, Director of Public Works, Fort Carson, Colorado

BACKGROUND

Fort Carson is a U.S. Army garrison mountain post established in 1942 just outside Colorado Springs, Colorado. At that time, it was an economic development effort of the city: to buy a ranch and dedicate it to the U.S. Army. Today, Fort Carson is the second largest employer in the state of Colorado, generating about $1.3 billion per year in the local economy.

Fort Carson is adjacent to the southern edge of the Colorado Springs metro area, which has a population of approximately 625,000 people and borders the installation on two sides (Figure E.1). Fort Carson directly supports about 25,000 soldiers and has about 150,000 people, including retirees, that come to the installation for various services.

Fort Carson is about 2 miles wide and about 5 miles in length from north to south. The installation includes an airfield and six operating gates through which about 90,000 trips are made each day, and the traffic is projected to continue growing as soldiers return from deployment in the next 4 years and the area's numerous military retirees, who access Fort Carson shopping and services, grow in number. In the past 5 years, Fort Carson's building square footage has increased by about 50 percent, due mainly to the growth of Fort Carson's soldier population. The garrison hosts about 10,000 soldiers—approximately 7,000 single soldiers in barracks and about 3,000 (eventually 3,500) married soldiers in on-post family housing.

The primary mission of the Mountain Post Garrison Team is to provide mission support and services including quality of life programs for Fort Carson soldiers, families, and community to enable forces to execute expeditionary operations and to minimize stress on soldiers and families in a time of persistent conflict.

Soldiers and families are under a great deal of stress. It is not uncommon at Fort Carson and other military installations to have soldiers who are on their fourth or fifth combat deployment. One of the best things that facilities managers can do to minimize stress is to create quality facilities for those soldiers and families. An added, long-term benefit is that it costs less to provide quality facilities because high-performance buildings save water and energy.

FIGURE E.1 Fort Carson and Colorado Springs.
SOURCE: Courtesy of Fort Carson.

Sustainable development—high-performance buildings that minimize sustainability impacts—provides many tangible and intangible benefits. Because its execution sometimes means additional up-front costs and/or design time, it is critical that Fort Carson and other military installations on the sustainability journey have the continued support of military and U.S. government leadership; the quality of life for the soldiers and their families, and by extension the quality of our military, depend on it.

SUSTAINABILITY GOALS

In 2002, Fort Carson was one of the first three U.S. Army installations to pilot the concept of sustainability. At that time, Mary J. Barber and Tom Warren of the Fort Carson Directorate of Environmental Compliance and Management, invited people from around the community and state to help Fort Carson personnel set 25-year sustainability goals.

The charge was: What will Fort Carson actually do and look like if it were to be sustainable? Following education about what actually becoming sustainable means, seven visionary performance goals (Box E.1) and five process goals were developed for achievement by 2027.

To achieve these goals, the installation began with a leading-edge hybrid management system that combined the aspirational sustainability goals with the U.S. Army- and U.S. government-required Environmental Management System (EMS) designed to ensure environmental legal compliance. The 25-year goals were managed using 5-year objectives and 2-year work plans; continued involvement of community stakeholders was encouraged but not required.

At present, the sustainability goals have been integrated into the garrison's strategic plans, which are updated by the garrison commander to reflect multiple objectives related to soldiers, families, and the workforce and Fort Carson's training mission. The EMS has returned to its traditional focus on environmental compliance. Annual EMS audits are conducted through a U.S. Army self-auditing system.

In addition to the 25-year goals that the garrison commander committed to in 2002, he committed Fort Carson to annual reporting back to the Colorado Springs metro community on goal progress. As the goals were set at a conference-like event, the Fort Carson sustainability team produced annual community-inclusive sustainability conferences beginning in 2003 to provide updates on Fort Carson's

---

BOX E.1
Fort Carson Sustainability Goals 2002-2027, as of 2010

- 100 percent renewable energy, maximum produced on the installation
- 75 percent reduction of potable water purchased 2002-2027
- Sustainable transportation achieved, characterized by 40 percent vehicle miles reduction from 2002 and development of sustainable transportation options
- Sustainable development (facilities planning)
- Zero waste (solid waste, hazardous air emissions, wastewater)
- 100 percent sustainable procurement
- Sustaining training lands—meaning ongoing capability of the biological health of training lands in support of the installation's primary mission to train soldiers

---

progress and to provide the metro region with sustainability education and inspiration. Typical conferences hosted more than 500 participants. These Fort Carson community sustainability conferences became important annual events for the entire sustainability community of the region; conferences typically included more than two dozen supporting non-profit partners, ranging from environmental and peace groups to chambers of commerce and educational institutions.

To maximize the effectiveness for the region's sustainability performance, which itself is a critical aspect of Fort Carson's sustainability success, Fort Carson transitioned the conference to being hosted by a community non-profit and established an educational effort for all of southern Colorado, called Southern Colorado Sustainable Communities. Fort Carson organizes a "military track" as one of three to four primary subject matter tracks at the conference. Sustainable Fort Carson, the newly adopted name of the program, continues to produce an annual sustainability progress report for public distribution.

Another important ongoing feature of Fort Carson's community stakeholder-based approach to sustainability performance is a monthly breakfast discussion of the achievements and challenges facing a specific sustainability goal, hosted by the garrison commander.

The sustainability education of the conference and the sustainability goal breakfasts are enhanced by a three-tiered approach to sustainability training for installation soldiers and civilian employees. Soldiers receive ongoing environmental compliance training and unit-embedded assistance from the Environmental Division of the Directorate of Public Works (DPW). The Environmental Division and the Sustainable Fort Carson programs collaborate to produce and deliver two levels of sustainability training: awareness training for all soldiers and employees and competence training for managers that includes integration of sustainability performance with the installation's strategic plans.

A new garrison commander comes onboard every 3 years, and the DPW is now working with the fourth garrison commander since the goals were set.

The installation developed a logo for Sustainable Fort Carson in 2010 in order to better reach the nearly 30,000 soldiers and civilian employees and approximately 120,000 other installation users with consistent and modern messaging techniques (Figure E.2). It is hoped that the logo will help encourage more people to think, "I want to be part of this brand." It reflects current Garrison Commander Robert F. McLaughlin's understanding that sustainability is a "state of mind."

FIGURE E.2 Fort Carson Sustainability Program logo.
SOURCE: Courtesy of Fort Carson.

Renewable Energy

The local utilities and vehicle fuel providers are not ever planning to deliver 100 percent renewable energy to Fort Carson, despite the installation's goal. Obtaining energy outside the existing system is unlikely to be cost-effective, although projects are continually investigated and evaluated. So the installation decided to collaborate with the Pikes Peak Sierra Club and local and statewide energy experts to create a regional sustainable energy plan, called the Pristine Energy Project. A Fort Carson sustainability planner with sufficient expertise, co-author of this appendix Christopher Juniper (contracted through Natural Capitalism Solutions), is co-leading the project with the president of the local Sierra Club and a small group of local energy experts. Technical support is being provided as needed on a volunteer basis by the National Renewable Energy Laboratory (NREL), the Governor's Energy Office, and two non-profits, the Southwest Energy Efficiency Project and the Colorado Renewable Energy Society.

The Pristine Energy Project's focus is on identifying the barriers between the providers of sustainable energy and the customers who want it, including Fort Carson and other military installations in the region. For example, new wind-powered electricity is being considered by the local utility for sale to customers like Fort Carson, but the retail cost to the installation has been prohibitive. The Pristine Energy Project will create a plan by early 2011 that will identify for public policy makers a path toward helping the buyers who want renewable energy now and in the near future to be able to buy it from the providers at a reasonable cost. The plan will also outline a pathway to mainstreaming sustainable energy for the entire Colorado Springs metro region by 2030, supporting goals to achieve sustainability performance adopted in October 2010 (see Box E.1) by the regional sustainability planning effort that was also strongly supported by Fort Carson's garrison commander and Sustainable Fort Carson planners.

Vince Guthrie, the utilities program manager in the DPW, is working on additional aspects of sustainable energy. Early in the installation's sustainability journey, he purchased some inexpensive renewable energy credits (RECs) through the Western Area Power Administration (WAPA) in order to establish Fort Carson immediately on the sustainable energy path. The RECs offset approximately 28 percent of Fort Carson's electricity from 2004 to 2009. The purchase was considered a "placeholder" while the installation worked on direct sources, such as wind, solar, and biomass. Through the work of the utilities program manager with the private sector and Colorado Springs Utilities, the city of Colorado Springs municipal utility that serves the installation with water and electricity, a 2-megawatt solar array was installed in 2008 (Figure E.3).

The array supplies a little more than 2 percent of the installation's demand. Over its 20-year life it is projected to save $500,000, since the fixed costs of solar will gradually become lower than the currently projected increases in utility costs. Guthrie is also working with the Front Range Energy Consortium—

FIGURE E.3 Two-megawatt solar array installed at Fort Carson, Colorado, in 2008. SOURCE: Courtesy of the U.S. Army.

five Air Force and U.S. Army military installations that are investigating a 50-megawatt concentrated solar installation on a U.S. Army chemical depot site in Pueblo, 40 miles from Fort Carson. NREL is conducting a technical assessment of the proposed project in collaboration with the Army Environmental Command that is due in 2011, after which the partners will determine how best to move forward.

Two other efforts are under consideration: purchasing approximately 20 megawatts of wind power through the local utility if it becomes affordable, and the development of a wood biomass co-generation (electricity and heat) facility by the private sector on the installation, partly displacing the existing and aging natural-gas-powered boilers.

Concerning vehicle fuels, the Fort Carson civilian fleet fuel stations pump only E85 fuel, which is an alcohol-based, alternative fuel. And the installation, through the Pristine Energy Project, will complete a life-cycle sustainability performance assessment of all potential vehicle fuels and energy sources, including batteries, with a regional focus, in early 2011.

Sustainable Development (Facility Planning)

Fort Carson's sustainable development goal holistically approaches three areas: land use/master planning, high-performance buildings, and storm water management (Figure E.4).

FIGURE E.4 (Left) Permeable Paver Project and (right) bioretention/stormwater management. SOURCE: Courtesy of the U.S. Army.

Fort Carson has completed 13 U.S. Green Building Council (USGBC) Leadership in Energy and Environmental Design (LEED) Silver and 14 LEED Gold certified buildings, which may be the most in the U.S. Army on any one installation (Figure E.5). That did not happen by accident. The U.S. Army Corps of Engineers, Omaha District Corps of Engineers, and its resident officers at Fort Carson are the garrison's partners in design and construction.

The U.S. Army requires its installations to develop buildings that could *potentially* be certified as meeting the LEED Silver ratings. However, Fort Carson's DPW did not believe that the potential for certification would be as effective as formal certification by the USGBC. The first certified LEED Gold building in the U.S. Army was opened up to the 1st Brigade of the 4th Infantry Division more than a year ago (Figure E.6).

FIGURE E.5 Examples of LEED Gold and Silver buildings at Fort Carson, Colorado. SOURCE: Courtesy of the U.S. Army and Army Corps of Engineers.

FIGURE E.6 LEED Gold Headquarters of the 4th Infantry Division, Fort Carson, Colorado. SOURCE: Courtesy of the U.S. Army Corps of Engineers.

The design-build contractor of the facility, Mortenson Construction, initially targeted achievement of LEED Silver certification. Mortenson decided on its own to work toward additional points so that one of its projects could be certified as LEED Gold. It now can claim to have designed and built the first LEED Gold building for the U.S. Army. One of Fort Carson's successful strategies is allowing private-sector design-build teams to "be all they can be" because they see high sustainability performance as a branding and marketing opportunity. And Fort Carson recognizes that the more sustainable the building's design, the better it will serve the Army over the building's lifetime.

One of the key parts to the installation's sustainable development success is the passion exhibited by people in key positions both within the Corps of Engineers Omaha District and within the Fort Carson DPW. It is critically important to have motivated and LEED-trained people in the right positions, creating excitement and passion about high-performance sustainable buildings.

Sustainable Transportation

The transportation challenges of Fort Carson begin with its being a rapidly expanding, spread-out installation at the edge of a sprawling metro area with a poor mass transit system—resulting in 93 percent of the people arriving at Fort Carson in single-occupant vehicles, which creates major bottlenecks at installation access gates as well as on-post congestion. The long-term sustainable transportation goal is to achieve a 40 percent reduction in vehicle traffic through demand-reduction strategies and the development of cost-effective and sustainable alternative systems. The objective by 2012 is a 25 percent reduction in single-occupant vehicles, with achievement of a 40 percent reduction by 2017.

The transit model in the community is broken, although long-term plans are in development for a new governance structure (a likely recommendation is a regional transit authority with an independent and stronger funding mechanism). When Fort Carson's sustainable transportation goals were established, public transit riders had to transfer just outside the installation's gate to get anywhere, and they had to transfer again to access the installation, unless using the mobility services for disabled passengers. It was 70 minutes between buses on Fort Carson and 70 minutes between the buses serving the three routes at that transfer station, so an 8-mile trip from downtown could take 1.5 hours.

Only about 20 or 30 riders a day used the system—because they had no other choice. The installation estimates that at least 50 to 100 transit-dependent people need to access the installation every day for employment or medical and other services. By 2012, Fort Carson aims to partner with transit-providing organizations in the metro area to create direct-to-installation services for 500 people that will attract enough people who have other choices ("choice" riders) that non-choice riders receive sufficient cost-effective services; the system is being designed to expand over time to serve 2,000 daily riders. With 2009 increases in the federal mass-transit benefit to $230 per month maximum (about $5 per commute), the potential exists for direct-to-installation services that require little if any local tax subsidies. Focus groups and polls tell DPW that if it takes more than 10 minutes to arrive at destinations by transit, choice riders will not use it. Collaboration with local transit providers to achieve these "stretch" goals is ongoing.

At present, transit buses achieve about 3 or 4 miles per gallon, meaning that a bus must have 11 people onboard to match the energy efficiency of today's cars (at the average occupancy of 1.6 passengers per vehicle). Buses will become even less comparatively fuel-efficient as cars such as the Nissan Leaf electric car, which is expected to achieve 99 miles per gallon equivalent, become available in 2011.

Ideally, the transit vehicles will be electric or electric-hybrids with maximum energy efficiency per passenger mile. Using electric power can reduce the $10 per hour fuel cost of traditional transit vehicles (of a typical $70 per hour total cost) down to about $1 per hour. The challenge is capitalizing the system with vehicles and supporting infrastructure, including the extra costs of electric vehicles.

Bus service to Fort Carson ended in 2010 because of city/regional tax revenue shortfalls. Federal government shuttle bus support is provided only for commuters or those engaged in Department of Defense business, or it can be provided by Fort Carson to its commuters only if all costs are covered by passenger fares. But as attainment of shopping, food, and medical services on the installation almost always requires motorized transportation, getting people to the gates without their cars only solves half the problem—with on-post carless mobility being the other half that is required to attract transit riders who could otherwise drive.

The installation's sustainable transportation team is researching a mobility system that does not use private-occupant vehicles and instead uses private-sector-provided car sharing, low-powered-vehicle sharing (bikes, electric bikes, or other personal mobility devices), on-call transit services, enhanced telework strategies, and expansion of pedestrian and low-impact vehicle infrastructure. In 2010, bike paths were added to main streets in the central cantonment area by transforming two-way streets into one-way couplets, which makes room for bicycle lanes in place of left-turn lanes, without roadway expansion.

What are some game-changing sustainable technologies? Fort Carson has closely examined the potential for a personal rapid transit (PRT) system. The first-phase study, completed in 2009, designed a PRT system for the installation and estimated its capital and operating costs and potential revenues, overall sustainability performance costs/benefits, and potential for private-sector development and operation. The second phase, to be completed in 2011, will design commuter extensions into adjacent communities, evaluate system sustainability performance compared to other future transportation alternatives, and provide private-sector development opportunities.

PRT systems provide the convenience of autos by means of a system of computer-guided "podcars" that travel on a separate right-of-way using electricity. A passenger or group of passengers (up to six per car) call for a podcar from a system station, which then takes them directly to their destination station without stopping. Travel speeds exceed those of light-rail and buses because of the direct station-to-station service and the minimal wait for a podcar when one is called. High energy-efficiency is achieved (about 10 times that of today's cars) because of the low weight of the podcars, their use of highly efficient electric motors, and the on-demand nature of the service such that vehicles are not moving unless demanded by a customer.

The preliminary Fort Carson-only system would cover 18 miles, deploy 36 stations, and cost $400 million to $500 million to construct. Over 40 years the cost per ride would be about $1.70, including all capital and maintenance costs—making it possible for a system that charges $2 per ride to earn a profit rather than require a subsidy. The system was designed for a podcar to be available to passengers within 30 seconds during non-peak hours and within a maximum of 8 minutes during peak hours.

A PRT system is now operating at Heathrow Airport in the United Kingdom (Figure E.7); the fundamental viability of the technology has been proven at a system operating in West Virginia for the past 30 years. Others are being planned in Sweden and in San Jose, California.

Another possibility is lithium-powered bicycles, which are now available for as little as $900, although typically they cost $2,000 or more (Figure E.8). The bikes will typically achieve speeds of about 20 miles per hour on their electric motors and will operate 20 to 30 miles on a single battery charge, depending on how much the rider pedals and other factors. With only half of a kilowatt-hour required to fully recharge them, lithium-powered bicycles get 2,000 to 3,000 miles per gallon equivalent. Bike sharing that uses electric bikes is better than traditional bikes on which riders may get hot and sweaty or overly cold in winter months. Private-sector companies are gearing up to provide electric-bike-sharing systems; Fort Carson is developing a request for proposals in FY 2011 for both bike-sharing and car-sharing on the installation.

Another sustainable transportation technology is the General Motors and Segway P.U.M.A. (Personal

FIGURE E.8 Electric bicycle.

FIGURE E.7 Personal Rapid Transit System operating at Heathrow Airport, United Kingdom. Photo courtesy of PRT Consulting. Inc.

FIGURE E.9 General Motors and Segway Personal Urban Mobility and Accessibility (P.U.M.A.) prototype. SOURCE: Courtesy of General Motors.

Urban Mobility and Accessibility) prototype vehicle (Figure E.9). The DPW is trying to determine if these low-impact vehicles, which operate on gyroscopes and are like a covered rickshaw that can transport two people, might be useful especially for carless mobility in the winter.

In short, Fort Carson is evaluating a wide range of technologies in order to identify and deploy the type of sustainable transportation system that would best serve installation users, would maximize life-cycle sustainability performance, and would be attractive to the private sector to build and operate. The system design will seek to maximize synergies between components, such as the multi-modal PRT station in Figure E.10 that is partially solar powered.

Sustainable Procurement

Fort Carson is pursuing adoption of an installation-wide sustainable procurement plan that will support compliance with Executive Order 13514 and the Department of Defense's Strategic Sustainability Performance Plan of August 2010. Annual progress toward more sustainable procurement is described in detail in Fort Carson's annual sustainability performance reports, available at the Sustainable Fort Carson Web site (www.carson.army.mil/paio/sustainability.html). The installation's sustainable procurement strategy includes life-cycle sustainability performance assessments of batteries, lighting, mattresses,

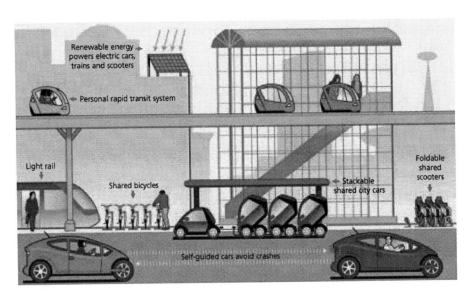

FIGURE E.10 Multi-modal, solar-powered Fort Carson personal rapid transit (PRT) station concept developed by PRT Consulting. SOURCE: Courtesy of PRT Consulting.

cleaning systems, laundry systems, vehicle fuels and energy sources, and transportation system options, to be completed and publicly available in fiscal year (FY) 2011.

At present, all new major construction projects at Fort Carson are being LEED certified through the USGBC. Contracts require the design-builder to submit an implementation plan that includes the following:

- An air quality plan,
- A waste management plan,
- A commissioning plan,
- A LEED schedule,
- A personnel role list,
- A 500-mile radius map to show where materials are going to come from, and
- A narrative on how every point to meet LEED requirements will be achieved.

This approach has energized construction-design teams from the very beginning to look at how they are going to create a LEED Silver facility, at a minimum, on Fort Carson.

Fort Carson's source selection boards look for contractors with past experience in LEED projects. Fort Carson's DPW has four LEED-accredited professionals on staff as of FY 2011. The U.S. Army, and maybe other agencies that use the LEED criteria, should consider this type of training.

F

Beyond Incrementalism:
The Case of Arlington, Virginia

Peter Garforth, Principal, Garforth International, LLC

I am going to talk about buildings as components of a community, as opposed to stand-alone energy objects. Rather than speak in generalities, I will present specific information regarding the development of a community energy plan (CEP) for Arlington County, Virginia, just across the Potomac River from Washington, D.C.—in many ways a suburb within the greater Washington, D.C. metro area.

The Arlington County process for developing a community energy plan kicked off in January 2010 with a workshop for a senior-level and very engaged community task force, whose ongoing role would be to guide the CEP process. The energy and greenhouse gas emissions baseline was already complete. In March 2010, a Community Energy Plan Technical Working Group (TWG) was formed. This group includes experts from both North America and Europe, allowing the TWG to look at the world from two different perspectives. The TWG subsequently met with representatives from a wide range of stakeholders, including county government departments, property developers, the Chamber of Commerce, local gas and electric utilities, Ronald Reagan Washington National Airport, and federal representatives from the Pentagon. The task force endorsed some tough transformative goals based on global benchmarking. A first town hall meeting was held in April 2010. The TWG's preliminary recommendations will be presented to the task force in September 2010, with a completion target for the CEP of March 2011.[1]

Arlington County's effort is a response to a strong belief that communities will need to manage energy strategically to remain competitive in the face of rapidly accelerating global demand for energy.

On present trends, by 2030 the world's energy demand could double from its 2000 levels. The newspaper headline for July 20, 2010, is that China is now a larger energy user than the United States.[2] However one looks at the issue of energy—whether environmentally, socially, economically, or from a foreign policy angle—it is a major challenge. In essence, we are looking at a future in which fears generated by volatile prices, uncertain availability, and climate change are colliding to create a growing

---

[1]The final draft of the Arlington County Community Energy and Sustainability Task Force report is available at http://www.arlingtonva.us/departments/DES-CEP/CommunityEnergyPlan/documents/file80565.pdf.

[2]See International Energy Agency, "China overtakes the United States to become world's largest energy consumer," July 20, 2010, at http://www.iea.org/index_info.asp?id=1479.

TABLE F.1 Energy Productivity Differentials

Region	Population (% of World)	Gross Domestic Product (% of World)	Energy (% of World)	Energy/Capita (Index)	Energy/Gross Domestic Product (Index)
United States	4.6	18.9	19.5	100	100
European Union	7.5	25.1	14.8	47	57
Japan	1.9	8.8	4.3	52	47
China	20.0	4.5	16.3	19	355
India	17.0	1.5	4.9	7	317
World	100.0	100.0	100.0	23	97

SOURCE: Data gathered by author from the International Energy Agency World Energy Statistics Web site (http://www.iea.org) and World Bank, World Development Indicators, 2007. Available at http://siteresources.worldbank.org/DATASTATISTICS/Resources/WDI07section1-intro.pdf.

public awareness of both the need for transformational changes around energy and the positive opportunities created.

The United States is spending roughly $1.5 trillion annually on energy. In terms of energy used to create a dollar of gross domestic product (GDP), if the index is 100 for the United States, the European Union is using 57 units of energy to generate the same dollar of GDP (Table F.1).

At a macro level, this gap represents a competitive disadvantage, or productivity opportunity, of about $600 billion annually. It represents a major opportunity for the United States to use innovation to close the gap or even leapfrog ahead of other major industrialized nations.

If we consider the three major sectors of energy use—industry, homes and buildings, and transportation—and index the performance of the United States on a unit basis against the performance of the European Union (they are comparably sized populations and economies: 310 million people in the United States, 494 million in the European Union, and a slightly bigger economy in Europe), then we see clearly different opportunities sector by sector (Table F.2).

Industry has generally done a good job of globalizing energy productivity. Table F.2 indicates that an aluminum plant in the United States and in Germany are using about the same amount of energy to make a ton of the same product. Industry seems to have already learned that global best practice is a requirement to remain competitive and has developed approaches to manage sharing. In transportation, there is a larger gap when measured per passenger-mile or per ton-mile. This gap is mainly due to heavier vehicles, less mass transit, and fewer high-performance diesels in the United States as compared to Europe.

The most interesting opportunity for significant energy productivity gains is the 40 percent of all North American energy used in buildings of all types and all ages, both residential and nonresidential.

TABLE F.2 Comparison of U.S. Energy Use to That of the European Union (EU), by Sector

Sector	Share of Energy Use	Index USA/EU (Indicative Ratio of U.S. Average to EU Average)
Industry	32 percent	1.2 : 1
Buildings	40 percent	2.5 : 1
Transportation	29 percent	1.4 : 1

SOURCE: Garforth International, LLC, estimates.

Even when adjusted for climate and service levels, this index is between two and three times higher in the United States than the average level in the European Union.

It is easy to forget how dysfunctional the energy supply chain is, especially when we look at it at a community level, not at a building level. Some round numbers: 60 to 70 percent of the potential energy of a fuel is used up getting the energy services to a building, whether those fuel sources are wind, coal, uranium, or wood. Then, we put the energy itself through the buildings into the building systems for heating, cooling, lighting, and appliances, such as the computer and the coffee machine. If we're lucky, we get 5 percent of the primary fuel calories as heat in the cup of coffee. So, that means we are basically spending $1.5 trillion to get about $150 billion to $200 billion worth of services.

So, what does all this have to do with the community? Well, the community is the lowest level at which all these things come together and where it is possible to influence the energy outcomes as an energy system spanning fuel to final service—something that is hard to do effectively on an individual building basis. The Arlington County CEP is being built around three goals: competitiveness, security, and environment. Competitiveness is measured by energy cost, employment, and inbound investment. Indicators for security include supply reliability, supply quality, and flexibility. Environmental impact is measured by the reduction of energy-related greenhouse gas emissions. The county is looking at these three issues as balanced necessities to retain both competitiveness and overall quality of life, while significantly contributing to mitigating wider environmental issues.

The 2007 baseline energy requirement of Arlington County was approximately 76 megawatt-hours-equivalent for every resident. Commercial buildings use half of the county's energy, and residential buildings are using about 25 percent. Half of the energy going into commercial and residential buildings is from the conversion losses in generating and transporting electricity incurred outside the community. The energy used to get electricity to the community accounts for the largest single portion of the total use.

Energy-related greenhouse gas emissions in Arlington County total about 14.6 metric tons per resident. That number includes emissions generated by the ground operations of Ronald Reagan Washington National Airport, the Pentagon, and Fort Myer, which are located within the county. And, again, non-residential or commercial buildings account for half of the carbon footprint; residential for about one-quarter. Transportation accounts for about 25 percent of the carbon footprint, of which 14 percent, is generated by visitors and only 11 percent by local residents.

At a national level the greenhouse gas emissions for the United States are at about 22 metric tons per person, and the European Union at about 10.5 metric tons per person. At a local level,[3] Arlington County is at 14.6 metric tons per person. Nearby Loudoun County, Virginia, generates about 14.2 metric tons of greenhouse gas emissions per person, and in 2009, set a challenging long-term goal of 6 metric tons per person. Similarly, the community in Arlington County has declared its support for breakthrough goals as opposed to a more incremental approach. The global benchmark is arguably that of the city of Copenhagen, with a carbon footprint of 3 metric tons per person, which has recently set the goal to become carbon neutral (Table F.3).

Community solutions require integration. Integration includes building, renovating, and operating both new and existing buildings at least 30 to 50 percent more efficiently than today's average building is built, renovated, and operated. Typically, the premium to reach these levels is no more than 1 or 2 percent of the design and construction costs. As an aside, my team recently calculated the cost for a 40 percent above-code improvement in energy performance and a 70 percent reduction in greenhouse

---

[3]Excludes energy-related emissions from aviation, maritime, most national defense, long-distance freight, land use, and forestry use changes and most heavy industry. These are included in the national figures.

TABLE F.3 Greenhouse Gas Emissions Indicators for Countries and Municipalities

Countries and Municipalities	Greenhouse Gas Emissions per Person per Year (metric tons)	Greenhouse Gas Emissions per Person Goals (metric tons)
Canada	22.6	N.A.
United States	21.7	N.A.
Denmark	14.1	N.A.
Germany	11.7	N.A.
European Union	10.5	N.A.
Arlington County, Virginia, U.S.	14.6	"Breakthrough"
Loudoun County, Virginia, U.S.	14.2	6.0
Guelph, Canada	12.2	5.0
Mannheim, Germany	6.0	7.5
Copenhagen, Denmark	3.0	0

SOURCE: Rough indicators, multiple sources. NOTE: N.A., for not available.

gas emissions on a project in Ohio.[4] We found it would cost an additional $1.50 per square foot on a development of 2,500 homes and 2 million square feet of non-residential space, which equates to a premium of about 1 percent.

There are other energy components that need to be integrated. If we look at Copenhagen, we see not only highly efficient buildings, but also widespread district energy systems that provide for district heating and cooling to most properties. District energy systems provide centrally managed supply and delivery of heating, cooling, and domestic hot water to many homes and buildings. Through a network of highly insulated pipes, district energy optimizes both the investments and efficiency of various heating and cooling supplies. A major benefit is the flexibility of fuel choice that it offers, allowing heating and cooling from both fossil and renewable energy sources to be easily and flexibly combined.

Even in Copenhagen, a significant portion of the electricity comes from coal, yet it is still the global benchmark for greenhouse gas emissions. This is largely due to the ability of the district energy system to use much of the heat that is typically wasted in more conventional systems. This underlines the basic truism that it is important to look at how efficiently we use resources, not just what form those resources take. As has been the case over many years in Arlington County, benchmark communities like Copenhagen have an urban design and transportation strategy designed to encourage walking and biking, efficient trams and trains, and ultimately city-wide infrastructure to support the wider use of electric vehicles. Copenhagen has a low unemployment rate, high-value employment, and solid inbound investment. It was recently voted the second most livable city in the world.[5] In other words, from an economic and competitive standpoint, it has not done Copenhagen any harm being green!

So, what are we looking at? In a sense we are looking at what we knew 25 to 30 years ago. All community energy plans should be prioritized top to bottom, following what is normally called the "loading order" (Box F.1).

The first priority is energy efficiency: Have you done everything that you possibly can to avoid needing the energy in the first place? This includes constructing and operating buildings as efficiently

---

[4]See the Dillin Corporation, Integrated Energy Master Plan for Planned Marina District, Toledo, Ohio, Perrysburg, Ohio, 2008. Report available upon request.

[5]"Travel Top 25," *Monocle*, Special Edition, December 15, 2009. Available at http://www.monocle.com/sections/edits/ Web-Articles/Top-25-Cities/.

BOX F.1
The Arlington Loading Order

1. Energy Efficiency—*If you don't need it, don't use it.*
   - Efficient buildings and vehicles
   - Urban design for transportation efficiency
   - Mixed use development for commuting efficiency
2. Heat Recovery—*If it's already there, use it.*
   - Use existing "waste" heat
   - Distributed combined power and heat
   - Plan commercial sites to maximize use of "waste" heat use
3. Renewable Energy—*If it makes sense, go carbon free.*
   - Renewable electricity—photovoltaic, wind
   - Renewable heat—solar thermal, biomass, geothermal
   - Renewable heat and power—waste-to-energy, biomass
4. Energy Distribution—*Invest where it makes sense.*
   - Flexible energy distribution (electric, gas, heating, cooling)
   - Accept multiple fuels and energy conversion technologies
   - Optimize local/regional energy choices

---

as possible, encouraging efficient vehicles, urban planning for transport efficiency, and ensuring that jobs are available locally to reduce unnecessary commuting. The second priority is heat recovery: If you already have a lot of heat, use it. One of the largest sources of incrementally carbon-free energy on Earth is the wasted heat side of the U.S. electric system. Local distributed generation, along with various heat recovery strategies, can put this and other waste heat to valuable use. The third priority is to move to renewable energy. If it makes sense, go carbon free, but only if you've done the first two things. And then the fourth priority: Team with existing networks to optimize investments between the systems. Do not fight the grid with high-priced renewables. Instead, work with the grid to optimize grid quality and things like avoidance of peaks. It's a classic model that requires an integrated approach among many community players for a sustained period of time. We tend to forget these priorities under daily pressures and end up with inefficient, sub-optimized urban energy systems.

So now we come to the community energy plan for Arlington County. From the outset the consultants encouraged the county leaders, who were part of the task force, to establish a framing target up front before embarking on the development of the plan. The CEP TWG needed to know whether the county wanted to develop a breakthrough plan or an incremental plan. If the plan was to be incremental, it would include community outreach supporting multiple efficiency and clean and renewable energy supply projects. Most of these efforts would focus on individual buildings, and generally would not rise to the level of the sometimes uncomfortable conversations where planning practices, local norms, and even policy may have to change. If a plan was to be transformative, it would need to look into scale projects that cover entire neighborhoods, where it is expected that there would be new norms, or even policies, created. As enough of these scale projects were implemented, they would ultimately define a new business as usual. It would take 20, 30, or even more years to get there, but it is important to know, before writing the plan, what the goal is.

A comment by one of the employees of Arlington County helped frame the conversation over transformation or incremental goals. When the then-county board chairman Jay Fisette (whom I should publicly thank for allowing me to show work-in-progress material) began to understand the kinds of new challenges that the CEP could represent, he commented, "This is not going to be easy." In response, the employee referred to said, "In Arlington, we don't do easy!" This became the backdrop catchphrase that helped lead to the task force directing the TWG to develop a plan to achieve "breakthrough targets."

Arlington County already has tremendous community activity around energy-efficient buildings and lifestyle and other aspects of sustainability such as recycling, water conservation, and urban tree cover. It is a very motivated community. I will not spend much time today on these admirable efforts. Strategically, these are measures that 10 years from now we probably won't even be thinking of as "green." In other words, the things that we already know we need to do to make a high-quality, efficiently built environment will be normal practice. Similarly, on the transportation side, many decades ago the county started to integrate transit planning into its urban design. It is arguably one of the best-integrated systems in the United States today, and by the way, it shows in the numbers of the actual energy performance and carbon footprint from the transportation sector.

It is one thing to plan for efficient buildings; it is quite another thing to actually deliver them. This brings us to energy performance validation, where there are real opportunities. All of us have heard the longevity statistics for buildings over and over again, but all too often we focus on the efficiency of new construction and overlook existing buildings. By 2040, nearly 90 percent of all of today's homes in Arlington County will still be standing. Sixty percent of the nonresidential property will still be standing. The county will also have gone through quite a bit of economic growth, adding yet more incoming workers during the day. If existing buildings are not renovated and operated to higher levels of efficiency, there is no way the breakthrough targets will be achieved.

Energy performance validation of all buildings is important for several reasons. First, transparent energy usage when buildings are bought or leased creates a market pull for efficiency, ultimately reflected in sales prices and rental values. Second, understanding ongoing energy performance tends to affect home and building operation practice, because users are more aware of the impact of their behavior. Third, energy performance validation is a potential marketing differentiator for new construction and major renovations. The CEP has the recommendation that most of Arlington County's buildings will have Energy Performance Labels. These labels represent a low-cost tool for validating actual energy performance. The Energy Performance Label would be available whenever a building is sold or rented. It would typically also be publicly displayed in buildings regularly used by the public. Actual energy performance would be independently certified. Both voluntary and mandatory approaches are possible; the county is opting for a voluntary program supported by the community at large. The specific labeling approach is not yet defined, but will probably use something like the emerging ASHRAE Energy Quotient approach,[6] which in turn is an adaptation of the European Union performance labels.[7] The schools and county buildings, along with some early adopters from the private sector, will be the first buildings to be labeled. Some of the property owners have already volunteered to be part of such a program. Again, this is a case of adopting and adapting a proven practice and using it for a number of things, including, potentially, as supporting documentation in a commercial transaction.

[6]American Society of Heating, Refrigerating, and Air-Conditioning Engineers, "ASHRAE Introduces Prototype of Building Energy Label at Annual Conference," June 22, 2009. Available at http://www.ashrae.org/pressroom/ detail/17194.

[7]EurActiv, "Energy Performance of Buildings Directive," September 29, 2010. Available at http://www.euractiv.com/en/energy-efficiency/ energy-performance-buildings-directive-linksdossier-188521.

Ultimately, though, many things have to be aligned to create successful implementation of community energy plans. Those that succeed have many common features, including the following:

- Leadership and community engagement,
- Transparency and outreach,
- Necessary planning policy changes in place,
- World-class energy efficiency,
- Integrated utility approach,
- Early implementation of "scale projects,"
- Magnet for business and academic excellence,
- Continuous improvement (raising the bar!), and
- Consistent execution over decades.

It is important to get serious quickly and take advantage of the benefits of scale. If you are going to do a military base, don't do one building, do the entire base. If you are going to do a home and there are 500 homes in the neighborhood that look the same, do all 500.

The TWG has identified potential scale project opportunities for early implementation, based on a number of criteria, including these:

- High probability of being implemented;
- Manageable number of participants;
- Large enough to implement integrated energy;
- Solutions within its boundaries;
- Possibility to apply different energy supply and efficiency than surrounding norms;
- Potentially economically, environmentally, and operationally attractive; and
- Future possibility to link to other community projects.

Typically, scale projects are high-density urban villages, core or downtown renewal, sport and recreation centers, academic campuses, military bases, and so on. The task force is looking for anything that, from an energy standpoint, looks like a small village, yet with relatively simple ownership or decision-making structures.

The CEP process has declared four high-priority areas in Arlington: Crystal City, which is adjacent to the Pentagon territory; Columbia Pike, which is a mixed-use development along one main highway; Rosslyn, which is a high-density, mixed-use urban environment with many buildings dating back to the 1960s that are rapidly going into refurbishment; and the transit-oriented development around the East Falls Church Metro station. Each one is different, with its own characteristics, and each one will be planned differently. But each one will have its own integrated energy master plan as a first step. The integrated energy master planning for Crystal City is in the initial development stage.

The energy use for Arlington County was modeled in 20 specific areas, each representative of the different kinds of neighborhoods. The sample areas represented 70 percent of the county's greenhouse gas emissions from buildings, 67 percent of energy use, and 35 percent of the land area. For each of the 20 areas, the consultants estimated the needs for heating, cooling, and other uses. From this the task force began to see the picture of the possibilities for sharing infrastructure, or where it could be possible to start sharing heating and cooling between buildings, ultimately leading to the development of neighborhood district energy systems. As building renovation proceeds, and where the energy density is sufficient, one can start inter-linking where it makes sense. Based on the analysis, about half of the county's energy use could be amenable to district energy.

There are potential benefits from district energy systems for just about everybody: the community, property users, property owners and developers, the district energy utility, and the incumbent gas and electric utilities. These are summarized below:

- *Community*
  - —Reduced environmental impact
  - —Increased supply security
  - —Fuel flexibility
  - —Possibility that peak load reduction may reduce tariff increases
  - —Migration of district energy (DE) benefits to lower density areas
  - —Investment opportunity in district energy utility
- *Property user*
  - —Equal or lower overall energy costs
  - —Less volatile energy costs
  - —Equal or greater supply quality
- *Property owner/developer*
  - —Reduced building investment
  - —Freeing of investment for enhanced efficiency
  - —More marketable space
  - —Reduced operations costs
  - —Investment opportunity in district energy utility
- *District energy utility*
  - —Profitable retail sales of heating and cooling
  - —Grid sales of clean and renewable electricity
  - —Fuel flexibility—extension to biofuels, fuel oil, and waste heat recovery
  - —Greenhouse gas credits
  - —Saleable expertise for similar projects elsewhere
- *Gas utility*
  - —Familiar business model
  - —Business diversification
  - —Higher sales volumes heating, cooling, grid sales
  - —Higher than average margins
  - —Greater knowledge of customers' requirements
- *Electric utility opportunity from district energy utility*
  - —Business diversification
  - —Low-carbon electricity to meet requirements
  - —Higher than utility margins
  - —Greater knowledge of customers' requirements

In the low-density areas of Arlington County where district energy is not feasible, the CEP calls for differential energy strategies based on energy efficiency, clean and renewable energy, local shared solutions, and transport efficiency. In other words, one completes the energy picture appropriate to the neighborhood, supported by activities that deliver results in line with the overall plan goals. The CEP for the county recognizes that "one size" does not fit all needs.

If done well, the CEP will create new business opportunities, reduce business risk, and increase the attractiveness of the community. The bottom line is that for the $1 billion to $1.5 billion that Arlington County spends every year on energy at today's prices, these efforts could reduce their total costs by $300 million to $500 million—revenues that can then be used for other purposes. Energy prices are unpredictable, but most observers expect them to rise in the coming decades, further increasing the value of successful implementation to the county.

I want to finish with one additional comment. When discussing these issues at the building level, we often say that productivity is the hidden benefit. We do all these incredibly detailed calculations on the energy bill, but in fact the real benefit is the productivity in efficient, well-managed buildings. The value of this productivity financially is far greater than the energy cost savings, which are not insignificant in themselves. The energy plan that the City of Guelph, Ontario, Canada, completed in 2007 became a planning role model for the entire country. We have recently heard that companies are investing in the City of Guelph as a direct result of its commitment to an integrated long-term approach to energy planning. This will bring many hundreds of new jobs to this city of 110,000 people. In other words, that approximately $300,000 which the city spent on the plan has already been handsomely repaid, even before it starts reaping the other benefits of the long-term implementation of the plan itself.

And one last comment on Copenhagen: The achievements there did not happen by accident, and it was not a historical peculiarity. In the early 1970s, Copenhagen had buildings of average efficiency and a relatively small downtown district steam system not unlike that of Indianapolis, San Francisco, or Manhattan today. Copenhagen made a conscious directional decision in 1973 to redesign the efficiency and energy system of the city. Today, it is a global benchmark of both efficiency and energy system integration. That was not an accident. That was a community decision: one that has been systematically and successfully implemented over decades.

So, I thank you very much, and I apologize for going through a wonderfully complex story quickly. I also want to publicly thank the people of Arlington County. I have never had so much fun working with a group of people. They keep us on our toes, but they never come in telling us why things cannot be done. They come in saying, "We want to understand it, but once we understand it, we see no reason why we can't get on with it and do it." And I think that is the absolute key to what we should be talking about today.

G

Getting to Net-Zero Energy:
NREL's Research Support Facility

Jeffrey M. Baker, Office of Energy Efficiency and Renewable Energy,
U.S. Department of Energy

Creating net-zero-energy buildings is a very challenging goal. However, getting to net-zero energy responsibly and affordably requires that projects achieve ultra-high energy efficiency first. By focusing on energy efficiency and taking advantage of what nature offers, ultra-high energy-efficient projects can be developed today using available technologies and acquisition techniques. The Department of Energy's (DOE's) Research Support Facility (RSF), located at the National Renewable Energy Laboratory (NREL), demonstrates what can be achieved with an unwavering focus on energy efficiency: a design that exceeds the benchmark ASHRAE 90.1–2004 energy performance standard by 50 percent. Indeed, my colleagues and I believe the RSF establishes a new energy performance standard that, if widely adopted, will help transform the energy performance of the nation's commercial building sector.

NATIONAL RENEWABLE ENERGY LABORATORY

The National Renewable Energy Laboratory is one of 17 national laboratories and major science capabilities operated by DOE. DOE is the single largest funder of physical sciences, with work performed not only in its national laboratories, but at more than 300 universities across the nation. Not surprisingly, DOE research and development funding has resulted in more than 80 Nobel Prizes—more than any other single research and development funding source in the world.

NREL is unique among national laboratories. While many of DOE's national laboratories have their genesis in the Cold War, NREL is the nation's only national laboratory to be created by public law. Originally named the Solar Energy Research Institute, NREL was created by Public Law 93-473, the Solar Energy Research Development and Demonstration Act of 1974, following the oil embargos. NREL's mission is to improve the nation's energy security, economic competitiveness, and environmental quality through research, development, demonstration, and deployment of energy efficiency and renewable energy technologies. NREL is located in Golden, Colorado, at the crossroads of the nation's energy industry (Figure G.1). NREL currently employs 2,300 scientists, engineers, and support staff and has an annual operating budget in excess of $350 million. DOE has designated NREL a federally

FIGURE G.1 Department of Energy's National Renewable Energy Laboratory campus in Golden Colorado. SOURCE: Courtesy of Pat Corkery.

funded research and development center and, as such, NREL supports DOE on a host of energy policy, technology, and market matters.

While NREL's beginnings were humble—its first buildings were mobile homes that were declared excess by the Bureau of Prisons—DOE's Office of Energy Efficiency and Renewable Energy (EERE) has since invested hundreds of millions of dollars developing its research infrastructure at NREL. EERE used this opportunity to develop NREL as an innovator in the field of energy-efficient commercial building design and construction. Starting in the 1990s DOE-sponsored capital construction has pushed the boundaries of energy efficiency in commercial buildings. The Solar Energy Research Facility, a 114,000 gross square feet (GSF) laboratory building completed in 1994, won numerous awards for its innovative and highly successful use of daylighting in office and laboratory space. The Science and Technology Facility, a 71,000 GSF laboratory building housing a highly complex research and development infrastructure for photovoltaic and related technologies, was the nation's first Leadership in Energy and Environmental Design (LEED) Platinum building. This focus on energy-efficient design of buildings at NREL, as well as NREL's site infrastructure, enabled the addition of almost 8 megawatts of renewable power generated from on-site wind and photovoltaic sources to provide power to the complex. Today approximately 32 percent of all of NREL's electricity needs are provided by on-site renewable production.

RESEARCH SUPPORT FACILITY

The Research Support Facility is a commercial office building and EERE's new corporate headquarters for NREL. The RSF was designed to be the most energy-efficient office building in the world and, as such, it redefines what is possible today in energy-efficient commercial building design. The RSF was designed to exceed the best energy performance standard available, the ASHRAE 90.1–2004 standard, by 50 percent at a cost that is comparable to similarly sized but less energy-efficient projects. In a testament to its energy efficiency, the RSF increases the square footage under roof at NREL by 60 percent but only increases site energy use by 6 percent. The RSF was completed in June 2010 and will house 825 employees in 222,000 square feet.

The shape of the RSF speaks to its function. The building is divided into two large wings to maximize the exposure to daylight, the heart of this highly energy-efficient design. Windows are plentiful, and the roof, which is covered with photovoltaic modules, is sloped to maximize sunlight exposure through the seasons. Combined with power-producing photovoltaic modules on surrounding structures, renewable electricity will provide the balance of electrical power required to operate the building. The completed RSF is shown in Figure G.2.

Achieving ultra-high energy efficiency did not require us to sacrifice building capabilities or comfort for energy performance. In fact, the integrated delivery, whole-building design approach, supported by extensive energy modeling, produced a largely passive design that met all of our mission requirements by using free environmental benefits such as ample daylight and cool, dry nighttime air. We estimate that the cost to achieve this level of energy efficiency is only 1 to 2 percent more than the total cost to design and construct a conventional office building on a square foot basis, with much of the additional cost attributable to the more intense and interactive design process. While not the best way to compare projects due the difficulty in obtaining "apples to apples" information, the cost per square foot of the RSF is $259. Based on a data set of 34 roughly similar projects captured by the Design-Build Institute of America's Design-Build database (www.dbia.org) and other publicly available sources ranging from LEED Certified to LEED Platinum performers, as well as projects with no LEED rating, 75 percent of these projects were more expensive than the RSF. This clearly demonstrates that highly energy efficient projects can be designed and delivered today at marketable costs.

The RSF's design was driven by a determined and continuous focus on energy performance and taking advantage of free energy provided by nature. The design standard for the project was set at 25,000 Btu per square foot—including plug loads. The final design, with allowance for a new corporate data center servicing the entire NREL site (not just the RSF), was about 33,000 Btu per square foot.

Avoiding any kind of lighting load was key to achieving energy efficiency, because lighting loads drive mechanical and other systems. Workspaces in this building are 100 percent day lit—by means of free energy and free light. To achieve that, the floor plate had to be fairly narrow, in this case, about 60 feet. Typically, to achieve this level of daylighting, the floor plate would be even narrower, at about

FIGURE G.2 The Department of Energy's Research Support Facility. SOURCE: Courtesy of the Department of Energy, National Renewable Energy Laboratory.

FIGURE G.3 Day lit interior spaces in the Research Support Facility. SOURCE: Courtesy of the Department of Energy, National Renewable Energy Laboratory.

FIGURE G.4 Labyrinth thermal storage at the Research Support Facility. SOURCE: Courtesy of Pat Corkery.

45 feet. The designer found a way to reflect light back into the building and make it day lit in a much wider floor plate, which was also much more efficient. On a typical day, even on an overcast day, there is virtually no need for anything but task lights (Figure G.3).

Reducing the lighting load through daylighting reduces, or in the case of the RSF, eliminates the need for traditional mechanical cooling. First-cost savings achieved through such an approach can be reinvested in other energy efficiency features such as the building facades, high-performing windows, and so forth. As the RSF demonstrates, this approach makes it possible to achieve ultra-high energy efficiency at a marketable cost per square foot.

The RSF walls are foot-thick concrete with an embedded layer of insulation. The walls not only keep the environment out, they serve as a giant thermal battery regulating heat gain or loss, allowing the building to operate at constant temperature, even in extreme weather. Building air supply is conditioned using stored thermal energy provided by an EERE-funded invention called a transpired solar collector in a concrete labyrinth in the RSF's basement (Figure G.4). Highly efficient radiant heating and cooling moving through 42 miles of tubing embedded in the concrete ceilings are used to heat or cool the building. Hot water is provided through a combination of renewable fuels and natural gas boilers, and chilled water is provided through evaporative cooling. The temperature outdoors was 100 degrees a couple of weeks ago, and I happened to be in the building and noticed someone walking by and shivering.

As I mentioned earlier, the energy budget for the building includes the site-wide data center and the computer system. To achieve this performance we realized that power management in an office building is absolutely critical. We replaced virtually all of our office equipment, including lights, phones, copiers, and so on, with more energy-efficient ones. Desktop computers were replaced with laptops, and we are working toward a thin-client solution that is even more efficient. Standard telephones were replaced with Voice-over-Internet. While most of our equipment was at the end of its service life and needed to be replaced anyway, the costs of doing so are not included in our cost per square foot for the RSF. All office equipment is monitored for activity, and if no activity is detected it is automatically shut off to save energy.

The windows are triple glazed and operable. We learned a lot about windows in this process and would do some things differently if we had it to do over, particularly in thermal breaks in the windows. About 60 percent of the windows open manually, with the balance opened automatically through the RSF's control system. During the nighttime, which is very cool and dry in Colorado, the windows are opened to purge the air in the building.

One of our obligations is to monitor how this building operates. The RSF is extensively metered, which allows us to show how energy use changes as operating conditions change and building components are modified. We are making it up a little bit as we go along, but we realize that we have to change things out, and people have to be able to test different components, such as windows. In this sense, it is a "living laboratory" that should generate data for many years.

Acquisition Strategy

Attention to the acquisition strategy was essential for the RSF's design and level of energy efficiency. The acquisition strategy was shaped by several factors.

First, we recognized that ultra-high energy efficiency required that the building's form and systems, occupants, and the environment needed to interact seamlessly. This recognition was important in setting design performance goals, that is, what was truly possible. Second was a determined and continuous focus on energy. Nothing was done in designing the building without first checking the energy models. Going back and checking every design decision against the energy model was critical, particularly when it was necessary to make trade-offs. Third was the fixed budget of $64 million, which had to include design, furniture costs, and everything else. Finally, as a national leader in the energy efficiency and renewable energy arena, it was EERE's obligation to push the boundaries.

Combining these factors led us to an integrated project delivery acquisition strategy, called a performance-based design-build, requiring the use of a whole-building design process. Although such an approach entailed a great deal of work, especially early on in the design process, it also created a great deal of value.

Our acquisition strategy focused on performance goals instead of the more traditional approach of providing technical specifications such as building size, construction materials to be used, and so on. We developed four overall performance goals: (1) an energy performance level of 25,000 Btu per square foot per year (exceed ASHRAE 90.1–2004 by 50 percent); (2) space to accommodate a staff of 800; (3) a design that would achieve a LEED Platinum rating; and (4) the fixed $64 million budget. That is essentially where we stopped with the specifications. This approach allowed the design team wide latitude to develop creative solutions to meet our needs.

Performance-Based Request for Proposals

The request for proposals (RFP) for this project was about 500 pages. It included three tiers of goals: "mission critical" goals that must be met in the first tier, "highly desirable" goals in the second tier, and "if possible" goals in the third tier (Box G.1). The only thing that we told the bidding contractors about these goals was that they were in rank order. We asked them to develop solutions that achieved as many goals as possible. This allowed them the freedom to work out the trade-offs themselves. (By the way,

---

BOX G.1
Request for Proposals Performance Goals for the Research Support Facility

Tier 1: Mission Critical Goals
- Attain safe work/design
- LEED Platinum
- Energy Star "Plus"

Tier 2: Highly Desirable Goals
- 800 staff capacity
- 25,000 Btu per square foot per year
- Architectural integrity
- Honor future staff needs
- Measurable ASHRAE 90.1–2004
- Support culture and amenities
- Expandable building
- Ergonomics
- Flexible workspace
- Support future technologies
- Documentation to produce "how to" manual
- Allow secure collaboration with visitors
- Completion by 2010

Tier 3: If Possible Goals
- Net-zero energy
- Most energy-efficient building in the world
- LEED Platinum Plus
- 50 percent better than ASHRAE 90.1–2004
- Visual displays of current energy efficiency
- Support public tours
- Achieve national and global recognition and awards
- Support personnel turnover

---

the RFP was modified eight times through collaboration between the design-build competitors in a bid to improve the end product and reduce risk to all parties.)

People ask, "Why didn't you just design it yourself?" After all, NREL has recognized experts in energy efficiency. The answer simply is: We do not design buildings. That is not our job. There are specialists out there that do that. Our responsibility was to define the goals clearly enough that the specialists could really develop a creative solution, which they did. It is very inexpensive to do all this, as long as it's planned and implemented up front.

National Design Competition

Ten groups submitted proposals and then we narrowed it down to three. We gave the draft RFP to the three final competitors because we did not know if we had hit the mark exactly. We used the draft RFP as a way to build trust and understanding with the design community. In the end, NREL was able to use a firm fixed-price contract because the contractors knew exactly what we wanted, which reduced their risk.

Design-Build Project Delivery Approach

The design-build project delivery approach, as opposed to the more traditional design-bid-build approach, creates a good deal of apprehension in some parts of the organization, such as the acquisition and project management organizations, as we used performance goals rather than technical specifications. Keys to making the design-build acquisition strategy work are up front and continuous owner commitment and involvement, clearly defined performance goals, substantiation criteria for these goals, and a firm fixed-price contract that shifts the performance and financial risk to the contractor.

Unlike the more traditional design-bid-build approach, a successful design-build project requires an extraordinary commitment by the owner to work with the design team early and continuously in the design process. If you cannot make that level of commitment, design-build will not work and we advise that the strategy not be used for project delivery

Tools

Many tools are now available to help in designing energy-efficient buildings. Energy modeling is very sophisticated, but it is not yet perfect. In fact, we had to improve our energy models for this particular building to give us the support that we needed to make the best energy decisions. Design charrettes are a great way to help define and fuel performance requirements. Bringing people from academia and industry and users of the building together to help define performance goals is critically important. It helps you understand the state of the industry so that you can then take advantage of that knowledge to develop the RFP.

The Design-Build Institute of America conducts a great training session. We brought them in for a week to teach our staff how to use a design-build approach effectively.

Finally, there is a significant role for and value in owner's representatives. Typically, owner's representatives have been used in the latter phases of a project to help ensure that the owner is getting what they need. The cultural shift from design-bid-build to design-build can be a challenge, and sometimes we slip back into our old cultural ways. Having an owner's representative in the front end of the process, however, can help ease this cultural shift.

Other Factors

To create an energy-efficient design, you have to take advantage of nature. In Colorado, we are blessed with dry air and lots of sunshine. The original design for the building had a two-wall system, but that evolved to a single-wall concrete system. The designers were trying to determine how to move energy around using the airspace available in the two-wall system. It was prohibitively expensive and was not going to work. The designers went back to search the Web for products that would meet the objectives using a different design. Lo and behold, they found transpired solar collectors, developed through EERE-sponsored research at NREL. These collectors are sheet metal panels with precisely designed and placed holes through which air is drawn. The pre-heated air is used to store heat in the RSF's labyrinth, which ultimately pre-heats the air used in the building's ventilation system at virtually no cost. It is an example of how the national laboratories had an impact on things. Twenty years after the transpired solar collectors were patented, they came back in a somewhat happenstance way to be an important element of the building (Figure G.5).

Other technologies and tools used in the RSF were developed or improved through EERE research at NREL, including a photovoltaic module for the production of electricity from sunlight, photochromic glass that darkens when heated, and photoelectric glass that darkens when a small electric current is applied, in order to shield occupants from direct sunlight, and, of course, the energy models that were critical to the RSF's design. All of these technologies are available today and, in a project that has been designed to be highly energy efficient, can be deployed affordably.

About 30 percent of a building's performance is attributable to the occupants and how they use the building. Occupants really make the building work or not work. In this particular building, the occupants cannot bring in coffee pots or space heaters. They have to make sure that they are conscious about how to use energy all of the time. To make sure that the building operates well, occupants need to be good citizens.

One of the biggest cultural challenges was furniture. The existing offices were set up in leased spaces with a lot of hard walls and private offices. To optimize the daylighting in this building, we knew we would have to use lower walls and cubicles. To overcome the cultural hurdles, we set up a test office and actually put people in it for a year and a half to make sure that things worked well. Through that process, we worked with the furniture manufacturers to improve the layout (Figure G.6).

FIGURE G.5 Transpired solar collector. SOURCE: Courtesy of the Department of Energy, National Renewable Energy Laboratory.

FIGURE G.6 Interior space layout in the Research Support Facility. SOURCE: Courtesy of the Department of Energy, National Renewable Energy Laboratory.

To maximize our LEED points we used locally available materials. In Colorado and Wyoming, lodgepole pines are being killed off by a pine bark beetle. The designers used the wood from the beetle kill as architectural accents in the building: The wood has a beautiful blue-grey tingeing caused by the fungus that the beetles carry and is now prized for cabinetmaking and the like.

About 78 percent of the construction waste was recycled. The aggregate material in the foundations and the slab all came from the old Stapleton Airport. Stapleton was decommissioned about 10 years ago, and the runways—high-quality concrete—were ground up for use. There were massive piles of this concrete available, and we used a lot of it to create the walls in the foundation and building.

To sum up, what we wanted, going back to the performance goals, was a building that would house 800 employees, be certified LEED Platinum, and use 50 percent less energy than ASHRAE at 90.1-2004, and stay within budget. What we got, through our performance-based design-build integrated project delivery approach and our commitment to working with the private sector to lower the project risk through superior project definition, was *every* "mission critical," "highly desirable," and "if possible" performance goal contained in the RFP. In doing so, the RSF demonstrates that, through superior energy performance based on an ultra-high energy efficient design, getting to net-zero energy responsibly and affordably is possible today.

Additional information about the Research Support Facility building, including the contractual documents, is available at two Web sites, http://www1.eere.energy.gov/buildings, and at http://www.nrel.gov/sustainable_nrel/rsf.html.

H

Sustainable Asset Management:
The Case of Los Angeles Community College District

Thomas L. Hall, Facilities Program Manager
Los Angeles Community College District

The Los Angeles Community College District (LACCD) is made up of nine colleges (Figure H.1) located throughout Los Angeles County, an area of 882 square miles, which includes 36 cities. The total student enrollment is about 250,000; most attend part time. Students are typically attending evening classes that are in session around 7:00 p.m., which offers a unique challenge in terms of using solar power as a source of renewable energy.

Eighty-two percent of the student population are from minority groups, 68 percent are immigrants or children of immigrants, 60 percent are female, 54 percent are the first generation in their family to attend college, and 40 percent are from households with incomes below the poverty line.

The rates charged by the colleges are relatively low, and the funds for sustainability efforts are limited. The district's operational funds are used to pay teachers and utility bills. So, a major challenge for the facilities management program is to leverage available resources and save money that can then be put back into teaching and the classrooms.

In 1978, California passed Proposition 13, which effectively shut off funding for school construction for more than 20 years. In 2000, Proposition 39 was passed, which permitted approval of school district bonds by a majority of 55 percent. Several bond measures were subsequently approved, and, along with some additional state and grant funding, brought the total school construction funding level to $6.2 billion.

When it was established, LACCD was part of a K-14 system in California, which means that community colleges were at one point treated as an extension of the secondary education system. Many of the structures used by the community colleges were bungalows or modular buildings. A decision was made to use the bond funding to construct buildings that would look like institutions of higher learning (bricks and mortar).

The first bond measure for $1.2 billion was passed in 2001. As the district board was discussing how the funds should be spent, some environmental groups came to the board and asked, "Now that you have all of this money, why don't you build in a sustainable way?" The board agreed that it was a good idea and adopted a sustainability policy in 2002. Shortly thereafter, it decided that all new buildings would be designed and constructed to meet Leadership in Energy and Environmental Design (LEED)

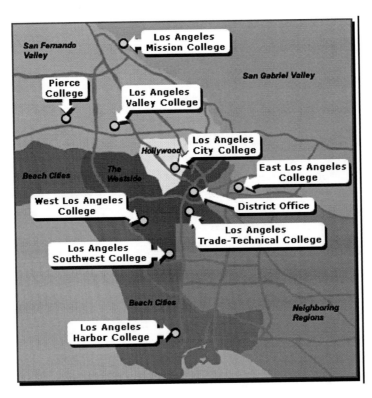

FIGURE H.1 Los Angeles Community College District. SOURCE: Courtesy of the Los Angeles Community College District.

certification standards[1] and that all new buildings would have a minimum of 10 percent on-site genera-tion of renewable energy with 15 percent to 25 percent of the building's energy being produced from renewable sources. As the program has been implemented, the LACCD is exceeding those standards.

The $6.2 billion in funding will pay for a total of 87 new major buildings and many retrofits (such as those at Mission College and Southwest College; Figure H.2). Thus far, 6 buildings have been con-structed and have been LEED certified (2 Gold, 3 Silver, and 1 Certified). Six additional buildings are currently in the certification process. Twenty-five buildings are under construction, 8 are in design, and there are more to go. Of the 87 new buildings, we anticipate that 25 will be certified as LEED Platinum.

Because buildings account for 39 percent of total energy usage for LACCD, there is a huge opportu-nity to save money through efficiency measures. Only 15 to 20 percent of the total cost of a building is for programming, design, and construction, while 80 to 85 percent is for operations over the life cycle. In a community college district where we have capital funds but limited operational funds, we want to ensure that we design and construct buildings that can be efficiently operated and maintained for many years.

As we started on the total construction program for all nine colleges, we began developing guidelines to provide some consistency and efficiency for the colleges and for the architects and engineers hired by the colleges. The guidelines focused on design that included water efficiency features, sustainable furniture, and carpets; cradle-to-cradle products; building information modeling (BIM); building site management (orientation); and an energy plan. The guidelines are posted on our Web site (http://www.build-laccd.org).

---

[1]See http://www.usgbc.org.

FIGURE H.2 Mission College parking garage (*left*) and Southwest College's maintenance and operations building (*right*). SOURCE: Courtesy of the Los Angeles Community College District.

FIGURE H.3 Pierce College botanic garden, reflecting Southern California's environment and the need for water efficiency. SOURCE: Courtesy of the Los Angeles Community College District.

LACCD has been working on water efficiency because Southern California is very dry (Figure H.3). We are considering many different elements and strategies, including low-flow irrigation, no irrigation and the use of native plants, cisterns for water capture, grey water systems, zero storm-water runoff, waterless urinals, and dual flush toilets.

One example of the measures that LACCD is taking in relation to design, materials and resources, and cradle-to-cradle products is related to sustainable carpet. The LACCD buys large volumes of carpet, and that creates enough of a market that it was able to specify the characteristics of the carpet and challenge the industry to produce it. LAACD hired an expert in carpets to help write the specifications, which were the following: carpet with a 30-year warranty, 100 percent recycled backing, 40 percent recycled yarn content, a wide range of colors and patterns, and a cost of $15 per yard. (Federal agencies also have this kind of buying power and could probably collaborate on specifications for sustainable products and challenge manufacturers to produce them.) The carpet mills actually changed the way they were manufacturing carpet in order to meet LACCD's specifications. The warranty is for 30 years as long

FIGURE H.4 Sustainable carpeting installed at Mission College. SOURCE: Courtesy of the Los Angeles Community College District.

FIGURE H.5 Conference room furniture available at Los Angeles Community College District's furniture procurement showroom. SOURCE: Courtesy of the Los Angeles Community College District.

as the carpet is maintained according to the manufacturer's specifications. At the end of 30 years, the carpet mills will take the carpet back and recycle it. In addition, the LACCD cut its carpet costs from $30 per yard to $15 per yard, leveraging both capital and operational funds (Figure H.4).

LACCD has also used master agreements to leverage its purchasing power and to provide consistency and efficiency for its colleges. For example, LACCD wanted to include the use of recycled material in its furniture. One challenge for public agencies is a lengthy bidding process. For LACCD, however, once a master agreement is in place, all of its colleges can use it, which saves time and money. In the centralized furniture procurement operation, LACCD has been able to specify furniture that contains 100 percent recyclable material, end-of-life return to the manufacturer, no use of chrome, a 15-year fully unlimited warranty, and below-market prices for design, delivery, and installation (Figure H.5).

LACCD has shifted its project delivery approach from the traditional design-bid-build, low-bid approach (which results in a lot of change orders, and one ends up spending 10 percent more than planned) to a design-build approach. Design-build project delivery requires that the owner become involved up front in the design process. LACCD is considering using integrated project delivery in conjunction with design-build. LACCD is also starting to talk about a design-build-operate approach that will focus on the entire life cycle of a building, not just on design and construction.

Design-build traditionally combines the design and the construction process, but it does not go quite far enough. LACCD has been talking to companies such as AutoCAD about their BIM systems, which we believe need to continue from design and construction into the operations stage, so that we can continue to use such systems.

Being in California, we are also using animation software in the design process. Many of our students go to work for Disney and similar companies, so we are using students to create a lot of the animation. They develop a flythrough view of the project as the modeling is taking place so that users and others can see what the building will look like before it is constructed.

We are just starting to explore BIM Storm, which is cloud computing for architecture. It brings together different types of software, which allows us to design, model, and construct a building in less time than is typical.

LACCD is also using several approaches to building commissioning. Commissioning to comply with the LEED rating system focuses on the heating, ventilation, and air conditioning (HVAC) system. In LACCD's last bond measures, funding for whole-building commissioning was included. LACCD has now moved to whole-building commissioning and is looking at everything throughout the building. Retro-commissioning is being used to ensure that mechanical systems in existing buildings are operating as designed. We are currently exploring the idea of ongoing commissioning, whereby LACCD staff would continually "tweak" a building's operating systems.

After the construction program started, LACCD realized it needed an energy plan. We understood that we needed to centralize the energy distribution, reduce the energy demand, and determine what sources of renewable energy could be put in place. We also knew that the long-term budget was a consideration in developing the implementation strategy. As a community college district, we also wanted to be able to transfer such knowledge to the students so that they can go out and replicate some of these approaches.

With regard to centralized distribution, LACCD is building a number of central plants, with differing characteristics. For example, the Valley College central plant incorporates solar thermal, which takes care of the heat and the cooling load of the campus by using an absorption chiller for cooling and stored hot water for heating (Figure H.6). To support centralized energy systems, we had to install extensive infrastructure for energy distribution (Figure H.7), and these costs had to be figured into LACCD's total budget and expenditures.

With regard to energy demand management, LACCD used a traditional energy-saving contract— ESCO (energy service company)—approach. Using ESCOs, we were able to retrofit lights, fans, pumps, and other energy-consuming components; install insulation, low-E glass, white and green roofs, and other conservation features; install state-of-the-art technologies, including occupancy sensors; and install metering and monitoring systems. The advantage of an ESCO is that the vendor pays for the energy-saving features up front and then the owner can pay back those costs over time through energy savings. Once the owner has paid back the amount of funding stipulated in the contract, the energy is free.

The renewable energy solutions considered by LACCD included solar, wind, and geothermal. Solar is probably the best answer for LACCD, given the Southern California location, but there is also potential for the use of wind and geothermal. Currently eight of the nine colleges have new solar arrays, including some thin film arrays (Figure H.8). The use of a solar concentrator is being evaluated for a couple of areas, but this technology has not yet been installed.

Regarding wind generation in urban areas, people do not want to see the big windmills in the middle

FIGURE H.6 Valley College central plant components: solar array (*top left*), hot water storage (*top right*), and vacuum tube heat-pipe collectors (*bottom*). SOURCE: Courtesy of the Los Angeles Community College District.

FIGURE H.7 New infrastructure for centralized energy distribution at Pierce College. SOURCE: Courtesy of the Los Angeles Community College District.

of town. Thus we are looking at devices that can be installed on a building's parapet and will probably install such devices on the parapets of some of our parking garages (Figure H.9).

Geothermal systems may come into play in reducing LACCD water usage. Our cooling towers use a lot of water. We may be able to use geothermal energy for cooling the water that comes back from the chill loop. We would install a number of wells on a campus. As one well gets saturated, we would move to another well.

Financing is also a component of the energy plan for LACCD. We have looked at a number of options, including the use of the federal energy tax credits, depreciation through third-party financing,

FIGURE H.8 Thin film solar array at East Los Angeles College. SOURCE: Courtesy of the Los Angeles Community College District.

utility incentives, renewable energy credits, carbon trading, bulk procurement, power purchase agreements, and lease agreements.

In order to take advantage of all the different incentive programs, like tax incentives, we have to get a third party involved that can capitalize the these programs and offer a discounted system. One of the biggest challenges with power purchase agreements is working with utility companies. Three of the colleges are in Southern California Edison territory, a public-owned utility territory where we have been able to purchase power from a third party. In a municipal-owned utility territory, however, this has been more challenging. Six of our colleges are in municipal-owned utility Los Angeles Department of Water and Power territory where the municipality has a city charter that says no one can sell energy in its district. A creative solution to this problem was a lease agreement whereby the LACCD's leases the solar equipment, and receives the benefit of the power that is generated. By using a lease agreement, LACCD still can take advantage of all of the different incentives.

I mentioned previously that most of LACCD's students are taking classes at 7:00 p.m. but electricity is generated during the daytime when the sun is shining, and so the issue of storage is an important one. LACCD needs to be able to shift its energy production to match its energy usage. To do this we are using thermal energy storage in the form of both ice storage (Figure H.10) and hot water storage.

We are also considering the use of lithium ion batteries for storage and will likely install a test system at the Trade Technical College within the next year. Other technologies under consideration include zinc-bromine batteries (hybrid flow batteries), vanadium redox batteries (flow batteries), and hydrogen storage (metal hydride).

FIGURE H.9 Wind turbine installation on the Van de Kamp Innovation Center. SOURCE: Courtesy of the Los Angeles Community College District.

As we look at LACCD's energy use, it is important to understand its utility bills. With regard to energy use, we are looking at lowering demand charges and changing time-of-day use. At one point we were looking at going off the utility grid, but it became apparent that we could use the utility grid to our advantage. Using the grid for energy storage is not out of the question, but we need to work out the details with the utility companies. There are several different scenarios. The first scenario is net metering—staying connected to the grid and running the utility meter back to zero. In this case, any extra

FIGURE H.10 Ice storage system at Southwest College. SOURCE: Courtesy of the Los Angeles Community College District.

energy produced is given to the utility, and you receive nothing from the utility company. The second scenario is virtual net metering—staying connected to the grid and running the meter back to zero. In this case, you get the same amount of energy from the utility company that you send to the utility. The third scenario is the feed-in-tariff—whereby the utility pays you for the energy you put on the grid at the time-of-day rate or a pre-negotiated rate, and you buy back whatever energy you need at the time of day that you need the energy. The feed-in-tariff offers the greatest potential for making renewable energies very affordable.

It is important to understand that LACCD is not in the utility business; we are in the business of educating tomorrow's workforce. For our situation, a key element is jobs, green jobs in particular, for our students. LACCD is using its campuses to develop curriculums covering all aspects of sustainability: technologies (solar, wind, geothermal); economics (business plans, life-cycle assessment); and operations and maintenance. The colleges are training for "green collar" jobs and for climate solutions today by offering courses, certificates, licenses, and degrees in sustainable areas of study.

LACCD has faced a number of challenges in developing sustainable campuses. One of the biggest is overcoming conventional thinking. People do not believe that you can do this. Another challenge is related to innovative versus proven technologies. We have found that banks are not as willing to finance innovative technologies. Further complicating the financing package is the use of third parties. In addition, in designing buildings to obtain LEED certification, people perceive "Platinum certified" to be very expensive.

Being an early innovator means that you run up against a lot of naysayers. You have to talk to a lot of people to convince them that a sustainable approach is possible and cost-effective, and you have to back up your arguments with facts. That takes a lot of time.

Another big challenge for LACCD has been that there are so many different user groups that we must interface with on an ongoing basis. Each one of our colleges and each one of our projects has a different user group. In order to deal with these circumstances, we had to develop some mechanisms to get many different people and groups to understand what we were doing. Constant communications and the development of sustainable standards and guidelines for design have helped us overcome some of these concerns. We have developed a motto of sorts, which is "Building a Green Tomorrow Today." Much more detailed information about the LACCD effort may be found at http://www.laccdbuildsgreen.org.

I

The Economics of Sustainability: The Business Case That Makes Itself

Greg Kats, President, Capital-E and Venture Partner, Good Energies

I'm going to talk about the cost effectiveness of "greening" buildings, drawing from my book *Greening Our Built World: Costs, Benefits, and Strategies*.[1] I wrote the book to address a fundamental question: How much does it cost to construct a green building compared to conventional buildings? Sponsors for the project included the largest real estate organizations in the country, the American Council on Renewable Energy, the American Institute of Architects, the American Public Health Association, Building Owners and Managers Association International, Enterprise Community Partners, the Federation of American Scientists, the National Association of State Energy Officials, the National Association of Realtors, the Real Estate Roundtable, the U.S. Green Building Council, and the World Green Building Council. The objective was to examine the issue from a balanced, in-depth perspective—greening the built environment is neither solely a nongovernmental organization initiative nor an environmental one.

We started with 350 buildings and worked with 100 architects, and by the time we were done, we were able to gather good data on about 170 buildings. We found that the perception is that building green costs about 17 percent more than building conventionally. However, the data show that the actual cost premium is closer to 2 percent of total design and construction costs, sometimes referred to as "first costs." This misperception of higher first cost seems to be very widespread. For example, I had the opportunity to go to Beijing last fall as part of the Obama administration trip. In China, the perception is that green buildings cost 28 percent more (Figure I.1).

The perception of higher cost seems to be the primary determinant for why people don't build green as a matter of course, which underscores the importance of gathering evidence-based data on this, communicating those data, and helping people understand that green buildings are an important step toward building more intelligently.

Figure I.2 shows data collected for utility bills, principally energy and water, for green office buildings. The additional cost of building green is about 2 percent, or $4 to $5 per square foot. If you assume that energy prices do not rise very fast, discount them at 7 percent, and assume only 20 years of operation

---

[1] G. Kats, *Greening Our Built World: Costs, Benefits, and Strategies,* Island Press, Washington, D.C., 2010.

135

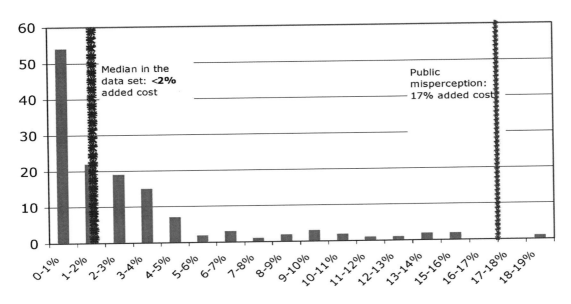

FIGURE I.1 Cost of building green: evidence from 146 green buildings. SOURCE: Greg Kats, Capital-E and Good Energies.

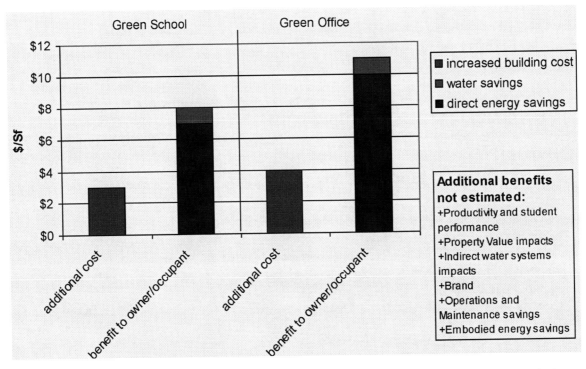

FIGURE I.2 Costs and benefits of green buildings: present value of 20 years of estimated impacts based on study data set collected from recent green buildings. SOURCE: Greg Kats, Capital-E and Good Energies.

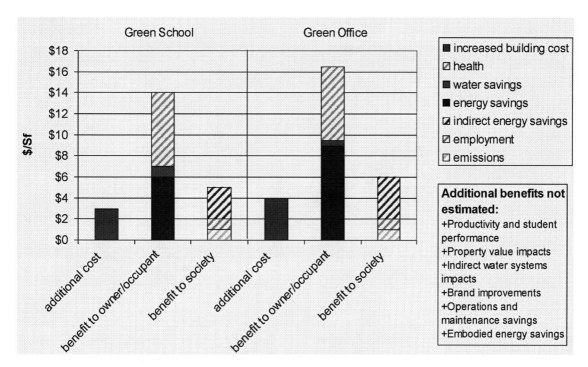

FIGURE I.3 Costs and benefits of green buildings: present value of 20 years of estimated impacts based on study data set and synthesis of relevant research. NOTE: There is significantly greater uncertainty, and less consensus, around methodologies for estimating health and societal benefits. SOURCE: Greg Kats, Capital-E and Good Energies.

(which is conservative because buildings clearly last more than 20 years), then the net present value from utility bill savings alone is almost three times greater than the first-cost design premium.

Thus, based on utilities alone, it is a fiscally prudent strategy to design and build green. Moreover, because there is a lot of uncertainty about energy and water costs—which are volatile and tend to rise faster than inflation—it is also a risk reduction strategy.

If you consider the larger set of benefits that accrue over 20 years—improved health, indirect energy savings, reduction of emissions, operations and maintenance savings, and so on—the savings add up. (When examining health-related issues, we relied particularly on the work that Vivian Loftness and her team at Carnegie Mellon University have done to compile and review hundreds of peer-reviewed studies.[2]) When you add up these benefits—the net present value of direct financial benefits primarily to the building owners, some to the occupants, and some to the community—the total benefits are about 10 times greater than the cost premium of constructing a green building (Figure I.3).

Over a period of 20 years, there are a number of additional benefits from green design relating to productivity, property value, and other factors. We were not able to quantify those benefits for this project, but we believe they are roughly the same order of magnitude as the benefits that we were able to quantify.

So, the question is no longer: Why would you design and construct a green building? It is instead: Why would you *not* design a green building? It is fiscally prudent to do so, and it entails lower risk. The next time someone says to you, We're thinking of designing a conventional building, you should

---

[2]Carnegie Mellon University, Center for Building Performance and Diagnostics. BIDS Tool. Additional information is available at http://cbpd.arc.cmu.edu/ebids/.

ask them, Who's your lawyer? I say this because the allergies, asthma, and respiratory problems associated with conventional design begin to have greater liability impacts when you can build green, much healthier buildings cost-effectively.

We also looked at 10 Midwest residential development projects with a combined total of 1,500 homes. In these projects, the homes were built in close proximity to each other, and 50 to 60 percent of the land was set aside for parks, walking areas, or trails. The site development costs per project were more than 20 percent lower on average. The costs per unit were about $12,000 less than conventional development, primarily due to lower infrastructure costs. In addition, the initial sale value was higher, and subsequent value appreciation was greater (Table I.1).

Green development is not only about individual buildings, but also about how buildings are located in relation to each other. The argument that you cannot build green without giving up economic benefits, at least for the building sector, is manifestly wrong. Interestingly, of the 170 buildings we studied, 18 were at least 50 percent more energy efficient and about one-third used some on-site renewable energy. The average CO_2 reduction for these 18 buildings was about 65 percent, even though the technology used was 5 years old. The average payback for these buildings with two-thirds reduction in CO_2 from operations was about five times the initial cost over 20 years (Figure I.4).

The lesson from this study is that we can reduce energy use to a much greater extent than we are typically doing today. The kind of vision that the General Services Administration is laying out, in terms of very deep reductions, is supported by what we know about the actual cost premiums of deep reductions.

Executive Order 13514, *Federal Leadership in Environmental, Energy, and Economic Performance,* sets somewhat ambitious goals for federal agencies, but it could go much farther. In my opinion, some goals are too weak and, in some cases, need both interim and long-term performance targets. This would help builders, architects, engineers, and constructors understand that there are goals that federal agencies,

TABLE I.1 Conservation Development: 20 to 30 Percent Reduced Development Costs over Conventional

Description	Conventional Sprawl Costs ($)	Conservation Cost ($)	$ Change	% Change
Grading Subtotal	1,425,418	947,142	478,276	−34
Roadway Subtotal	2,313,896	1,512,412	801,484	−35
Storm Sewer Subtotal	1,145,639	519,544	626,095	−55
Sanitary Subtotal	1,502,840	1,105,282	397,558	−26
Watermain Subtotal	1,657,739	1,233,850	423,889	−26
Erosion Control Subtotal	35,684	35,684	0	0
Offsite Sanitary Subtotal	26,250	26,250	0	0
Landscape/Restoration Subtotal	284,200	665,192	−380,992	134
Amenities Subtotal	999,222	732,240	266,982	−27
Contingencies/Engineering/ Legal Subtotal (25%)	2,347,722	2,347,722	0	0
TOTALS FOR PROJECT	11,738,610	9,125,318	2,613,291	−22
TOTALS PER UNIT	38,237	26,839	11,397	−30

SOURCE: Greg Kats, Capital-E and Good Energies.

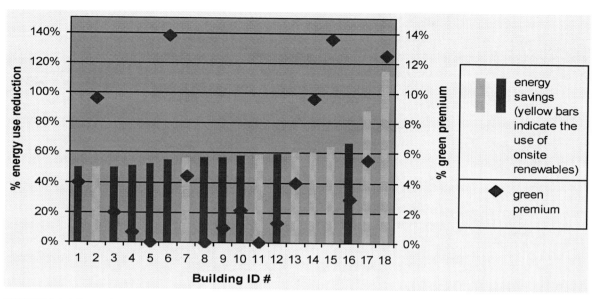

FIGURE I.4 Advanced energy savings and green premium: 18 buildings from the study data set. SOURCE: Greg Kats, Capital-E and Good Energies.

and in turn they themselves, have to respond to and that the goals become higher over specific periods of time. Executive Order 13514 requires the diversion of 50 percent of all construction and demolition waste by 2015. However, the average green building diverts more than 80 percent of construction and demolition waste cost-effectively today. So, why isn't there an 80 percent minimum mandate in this executive order? Similarly, zero-net-energy buildings by 2030 is a good goal, but we need interim goals, such as 50 percent lower energy use by 2018 and 75 percent lower use by 2025. Executive Order 13514 also calls for paper to include 30 percent recycled content. In my office and in my home, we use only 100-percent-recycled-content paper. So, why wouldn't the federal government establish a goal of 50 percent recycled content by 2015 and a goal of 80 percent recycled content by 2018? Setting such goals for federal agencies would signal to the market that there will be a large emerging demand over a finite timeframe, and then the market could build the capacity to respond to that market.

Proponents of green design are sometimes accused of promoting things that are only plausible for the wealthy or for the government. On the topic of green affordable housing, I had the good fortune of being the principal advisor in developing the Green Communities Criteria, which is now the national standard for design of green, affordable housing,[3] with 20,000 units built. The design and construction cost premium is about 3 percent, but the utility bills for these units are about 35 percent lower than those for conventional units. These units also show substantial improvements in indoor environmental quality. If we can build green affordable housing cost-effectively, then there is no building type that we cannot green cost-effectively. In my opinion, all of the Department of Housing and Urban Development (HUD) homes (and keep in mind that HUD spends almost $5 billion a year on energy bills) and leased buildings should follow the Green Communities Criteria. (I should add that HUD in this administration is doing a lot of green, healthy cost-effective design changes and programs already.)

There are other opportunities that could be mandated by an executive order. For example, greater coordination with the European Union (EU), California, and Massachusetts, which mandate zero-net-

---

[3] Available at http://www.practitionerresources.org/cache/documents/666/66641.pdf.

energy residential by 2020, while the EU mandate is for 2019. In addition, all new or retrofitted federal buildings should achieve a LEED Gold rating and reduce their energy use by 50 percent by 2015 and by 65 percent by 2018. Currently, there is public funding for building upgrades, such as lighting, with 2-year paybacks. But if you do a shallow retrofit, you can't go in and do a more serious energy efficiency upgrade. "Cream skimming" should not be allowed—that is, there should be no federal funding, subsidies, or tax benefits for retrofits that do not achieve either at least a 30 percent reduction in energy and water use or an Energy Star score of at least 90. The value of greening goes beyond energy savings.

Figure I.5 shows the Comcast Building, owned by Liberty Property Trust—a real estate investment trust in Philadelphia. Like many cities, Philadelphia is suffering from out-migration. Liberty built a super-green building; it is the tallest building between New York and Dallas. When reporters from the *Philadelphia Enquirer* saw the plans for this, they said, "This building challenges Philadelphia to be great again." So, it's not just about buildings. It's also about brand.

I think about brand as really three aspects. One is increased brand awareness. So, an owner of a new green branch bank is going to get a lot of positive free media coverage that drives traffic to the site. There are attribute-specific preferences—I might have health concerns and care about indoor environmental quality improvements, or I might live in Arizona and care about reduced water use. These specific attributes that I care about drive me to that building as a purchaser or tenant or client.

But, I think the largest brand-related driver here is non-attribute-specific preference—for example, the sense that it's a higher-quality building, which contributes to the perception that my brand quality is better. The LEED green design process is a more rigorous and integrated one, which results in a building that is more likely to be designed and built as intended and operated as designed. You reduce your risk and increase performance. That is why at least half of the corporate 500 firms that are building headquarters now build green: it's their face to the world. So, this larger brand aspect is hard to quantify, but ultimately it may be perhaps the largest driver in promoting green buildings.

We are also starting to see a significant premium for green buildings in terms of increased rental rates,

FIGURE I.5 Comcast Building, Liberty Property Trust.

1st Quarter 2008	Non-LEED	LEED Certified Offices	Difference	% Change
Occupancy rates	88%	92%	4%	5%
Rent ($/SF)	$31	$42	$11	35%
Property value ($/SF)	$267	$438	$171	64%

1st Quarter 2008	Non-Energy star	Energy Star Offices	Difference	% Change
Occupancy Rates	88%	92%	4%	5%
Rent ($/SF)	$28	$31	$3	11%
Sale Price ($/SF)	$227	$288	$61	27%

FIGURE I.6 Green building benefits: increased rent, sales, and occupancy. SOURCE: J. Spivey, "Commercial Real Estate and the Environment," CoStar, 2008. Available at http://www.costar.com/uploadedFiles/Partners/CoStar-Green-Study.pdf.

sales, and occupancy (Figure I.6). The premium for green buildings is about two and a half times greater for LEED Certified and Energy Star buildings than for conventional ones. So, again, green design is not only about higher financial return; it is also about risk reduction. In a buyers' market, people exercise their preferences, and they are starting to do so around green elements.

We know climate change is happening. As with smoking's link to cancer, the science is unambiguous about climate accelerating damage and costs. There are still perhaps 2 percent of climatologists who do not share this view—perhaps about the same percent of epidemiologists who do not accept the scientific consensus that smoking results in cancer.

The question is, What are we going to do about it? My company, Good Energies, a venture capital firm, is one of the largest investors in clean energy technology. It's a multi-billion dollar firm. I lead our investments in energy-efficient and renewable technologies. I wanted to mention a couple of these technologies because they represent the kind of technologies that can cost-effectively drive deep reductions in CO_2 emissions.

Figure I.7 is one example of technology about which we are excited. It's called "SageGlass Electrochromics." It allows you to vary the sunlight coming through a window between 2 percent and 65 percent. By itself, it can reduce the air conditioning load in a commercial building, on average, about 15 percent and peak about 25 percent. And, we're just scaling manufacturing that. There are a couple hundred installations.

I served as the director of financing for efficiency and renewables in the Department of Energy (DOE) for the Clinton administration, and early DOE support for this technology illustrates the kind of impact that federal support for research and development of fundamental technology can have. Figure I.8 illustrates recent work by the Federal Energy Regulatory Commission (FERC) which suggests that full deployment of distributed response (basically, demand management intelligent grid technology), could allow electricity growth to flatten from 1.7 percent down to zero. We have two smart grid investments in AlertMe and Tendril (I am on the board of both). Both are growing very rapidly. They allow

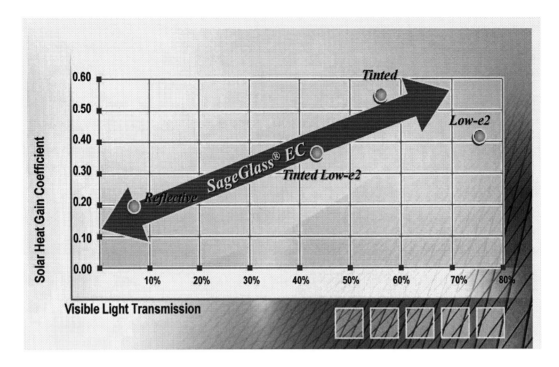

FIGURE I.7 Performance comparison.

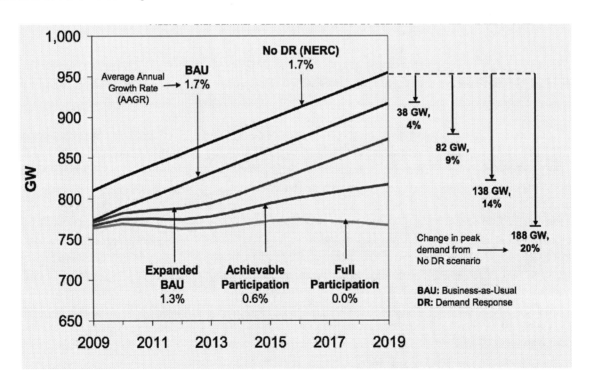

FIGURE I.8 Federal Energy Regulatory Commission (FERC) report: demand response potential. SOURCE: FERC, *Assessment of Demand Response and Advanced Metering,* 2008: assumptions: smart meters, dynamic pricing default, enabling technologies. Available at www.ferc.gov/legal/staff-reports/demand-response.pdf

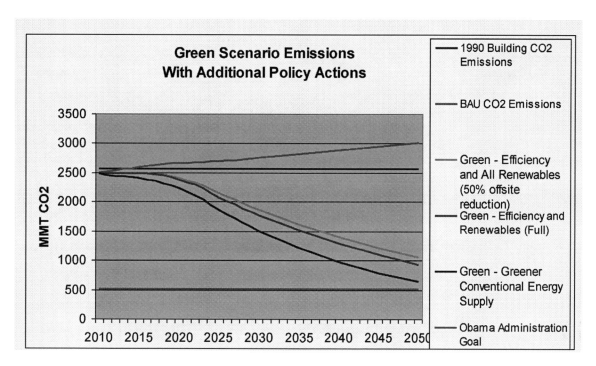

FIGURE I.9 CO$_2$ impact. SOURCE: Greg Kats, Cap-E and Good Energies.

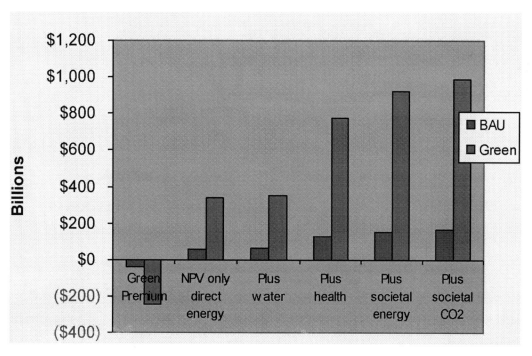

FIGURE I.10 Greening = wealth and jobs creation. Net present value (NPV) of net benefits of business as usual (BAU) and green. SOURCE: Greg Kats, Cap-E and Good Energies.

us to integrate across the meter, with a combination of efficiency and renewables, and drive toward this vision of deep reductions in energy waste through improved controls and enhancing comfort.

Let me turn, finally, to CO_2 emissions and the debate about whether mandating deep reductions in CO_2 emissions will hurt the economy. We modeled the CO_2 emissions from buildings under a range of scenarios and policy options (Figure I.9).

Then the question is, Does significant CO_2 reduction hurt the economy or not? Well, there is an up-front cost premium associated with greening all of those buildings. However, the direct energy savings resulting from green buildings creates about $350 billion in current value to society. Once you add in other direct benefits, the value creation is about $1 trillion in net present value, if you pursue an aggressive strategy toward green (Figure I.10).

So, although some may argue about climate change, the data are unambiguous: We can achieve very deep reductions in CO_2 emissions through thoughtful design, we can do it today, and we can do it cost-effectively. In my opinion, those who argue that we cannot are essentially saying that America has lost its capacity for innovation, that America has lost its capacity to drive through its political systems intelligent choices and the right regulatory structure, that America has lost its will to lead. I believe these pessimists are wrong. I think the investments being made by the federal government and the private sector will allow us to achieve deeper reductions and do so more and more cost-effectively.